2nd Fully Revised and Enlarged Edition

OBJECTIVE AGRICULTURE STATISTICS

NIPA GENX ELECTRONIC RESOURCES & SOLUTIONS P. LTD.
New Delhi-110 034

About the Author

Dr. K.S. Kushwaha : He was an intelligent student and secured first rank in the "District Board Exam" (8th Class) in 1969 and achieved 5th rank on the basis of state level merit of U.P. State prepared during 1969. He was awarded a fellowship Rs.100/- month during High School and Rs.150/- month during Intermediate classes of his study. He was asked by the Honourable Governor, Dr. M. Chennareddy to take admission in Govt.'s Queen's Intermediate College, Varanasi. He passed High School Exams with honour and obtained B.Sc. degree (1975) and M.Sc. (Statistics) in 1977 from Science College, B.H.U., Varanasi in first division.

Dr. Kushwaha completed one year "Professional Statistician Certificate Course" (P.S.C.C.) from Indian Agricultural Statistics Research Institute, New Delhi in 1979. He worked as "Research Fellow" for six months with fellowship Rs.1000/ - month in "Ford Foundation" New Delhi and completed his Ph.D. degree in 1991 from "Indian School of Mines, Dhanbad" Bihar in "Sampling Theory" with 12 research papers published in single authorship in Journal of Indian Society of Agricultural Statistics, New Delhi. For his continuous works he was awarded with prestigious "Dr. Radhakrishanan Award" on 4th Sept., 1992 by M.P. State U.G.C. and was declared as "The Man of Physical Sciences".

His papers have been referred in various text and reference books published by national and international publishers. He beard the responsibility of chairmanship in "Mathematics & Statistics Session" set up to evaluate the oral presentation of young scientists up to 28 years of age in scientific conference sponsored by MAPCOST, Bhopal and hosted by J.N.K.V.V., during Feb. 27-28, 1993 to give them "Young Scientist Award" at state level. His lecture in workshop On "Frontiers of Research in Finite Population Inferences" during (Nov. 1-9, 1993) financed by Govt. of India secured him second rank.

Dr. Kushwaha is an appointed member of (i) Executive Board and (ii) Editorial Board of "Bulletin Sample Survey Research Group, India", "Academic Council of R.D. University, Jabalpur," to guide Ph.D. Thesis of Scholars. He has published 19 research papers in the journals of National & International repute.

Dr Kushwaha retired as a Senior Associate Professor (Statistics) in the Department of Mathematics and Statistics, Jawaharlal Nehru Krishi Vishwavidayalaya, Jabalpur, M.P., India.

2nd Fully Revised and Enlarged Edition

Objective Agriculture Statistics

K.S. Kushwaha
Retired Head (Statistics)
M.Sc., P.S.C.C., Ph.D. (Statistics)
Recipient of Dr. Radha Krishnan Award (1992)
Associate Professor & Head (Statistics)
Department of Mathematics & Statistics
College of Agriculture, J.N.K.V.V
Jabalpur, M.P. (India)

NIPA GENX ELECTRONIC RESOURCES & SOLUTIONS P. LTD.
New Delhi-110 034

NIPA GENX ELECTRONIC RESOURCES & SOLUTIONS P. LTD.

101,103, Vikas Surya Plaza, CU Block
L.S.C.Market, Pitam Pura, New Delhi-110 034
Ph : +91 11 27341616, 27341717, 27341718
E-mail:newindiapublishingagency@gmail.com
www: www.nipabooks.com

For customer assistance, please contact
Phone: + 91-11-27 34 17 17 Fax: + 91-11- 27 34 16 16
E-Mail: feedbacks@nipabooks.com

ISBN: 978-93-91383-51-0

Composed and Designed by NIPA.

Preface to the First Edition

The book entitled **"Objective Agriculture Statistics"** has been designed for all P.G. Students of "Pure Statistics, Agricultural Statistics, Biological & Social Sciences" and those who have to appear in competitive examinations of I.S.S., S.S.S., I.A.S., State's P.S.C.'s. This book is useful for faculties of "Department of statistics" of Indian Universities. The book is the outcome of 26 years of teaching experience of U.G., P.G. and Ph. D. students of different disciplines of Agriculture, Agil. Engg. and Agril. statistics. in J.N.K.V.V. Jabalpur. The content of the book covers the syllabus on the topics "The Theory of Sample Survey Design, Designs of Experiment, ANOVA, ANCOVA Techniques, Transformation of Original Data and Non Parametric Methods."

The book contains 19 chapters, out of which chapters (1 - 8) deal with "The Theory of Sample Survey Design", the chapters (9 - 17) deal with "Designs of Experiments" and chapters (18 & 19) deal with "Transformation & Non Parametric Methods". In each chapter, three types of question " True/False, Fill in the Blanks and Multiple choice questions" alongwith the key answers have been provided.

During the preparation of the manuscript of this book, the author has incorporated the fruitful academic suggestions provided by Dr. Housals P. Singh, Prof & Head, School of Studies in Statistics, Vikram University, Ujjain, and Dr. Kamlesh Singh, Prof & Head, Department of Statistics, H.P.K.V.V. Palampur, Himachal Pradesh to enrich the book. It is expected to have a good popularity due to its usefulness amongst students, its readers and users.

The author is very much thankful and extends his honourable honour of higher order to Prof. Kalloo Gautam, the Vice-chancellar (J.N.K.V.V Jabalpur) for his kind permission to hand over the manuscript to " New India Publishing Agency" New Delhi to publish it in the present shape of a book.

The author is very much thankful and extends his honour to his mother - in - law Smt. Jai Sri Maurya for extending her motherly & caring attitudes to get rid of his bad health condition during (Nov-2006 to Feb-2007) at Rajajee Puram. Lucknow.

Sh. Madan Gopal Srivastava, the "Account Officer" of C.A.E, Jabalpur is duly acknowledged for providing his moral support and constant encouragement since 1988 and onward for the Ph. D. work and book writing programme of the author.

The author acknowledges very strongly to his wife Smt. Veena Kushwaha, daughter. Miss Akansha Kushwaha and son Mr. Utkarsh Kushwaha who cared a lot during the preparation of this manuscript.

Last but not the least, the author is very much thankful to the entire staff of "New India Publishing Agency," New Delhi for taking keen interest to publish the manuscript in its present shape.

Although the author has tried his best in designing the subject matter very nicely but that can be further improved based on the suggestions received from its readers & users. Therefore, the fruitful academic suggestions received from all sides will always be welcomed and highly appreciated by the author.

5th November, 2010 **K.S. KUSHWAHA**

Jabalpur

Preface to the Second Edition

Some additional topics on "Sample Survey" are attached with this book as "Supplementary Part". The additional topies are (I) Sampling on Successive occassions, (II) Non-Sampling Errors, (III) National Sample Surveys and (IV) Agricultural Statistic system in India.

It is hoped that this revised and enlarge version of the book will provide better coverage to those students who have to appear in various competetive examinations as cited in first edition and the "Faculty members" of the "Department of Statistics" set up in various central & state universities of Indian nation.

25 January, 2022 **(K.S. Kushwaha)**
Lucknow

Contents

1

Sample Survey Theory

1.1 Preliminaries

In this chapter some preliminary terminologies have been provided which are very much useful and frequently used in dealing with the theory of sample surveys and its application. In this chapter, the subject matter has been provided in three categories viz.

i) True/ False,

ii) Fill in the blank and

iii) Multiple choice question along with their answers.

1.2 True / False

Q.1 A function of all the sampling units of the sample drawn from the population, is known as population parameter.

Q.2 Random number method of sample selection is considered better than the reminder method.

Q.3 Collection of all the sampling units in the population in known as the sample.

Q.4 Sample survey has greater scope than complete enumeration.

Q.5 The fraction $f = \frac{n}{N}$ (where n is the sample size and N is the population size) is known as sampling fraction or inclusion probability.

Q.6 The precision of an estimator is directly proportional to its sampling variance.

Q.7 A sample survey may not produce more accurate results than complete enumeration.

Q.8 An estimator is a random variable.

Q.9 The ratio $f = \frac{n}{N}$, of sample size n and population size N is known as sampling frame.

Q.10 Collectivity (totalty) of sampling units is known as the population.

Q.11 Sampling units may be either individuals or group of individuals.

Q.12 $s^2 = \frac{1}{n-1}\Sigma(x_i - \overline{x}_n)^2$ is known as statistics.

Q.13 A list of all sampling units with their identification is not known as sampling frame.

Q.14 A complete enumeration is better than sample surveys.

Q.15 Census has greater scope than sample surveys.

Q.16 A function of all sampling units in the population, is known as an estimator.

Q.17 The sample estimator and sample estimate both have the same meaning.

Q.18 Enumeration of a small part of the population is known as complete enumeration or census.

Q.19 One of the roles of the sampling theory is to provide an estimate of sampling variance or M.S.E. of various estimators under consideration.

Q.20 A sampling procedure which depends upon the personal judgement or discretion of the sampler is known as "subjective sampling".

Q.21 A sampling procedure which is governed by a sampling rule or is independent of sampler's discretion, is known as "objective sampling".

Q.22 A sampling procedure in which every unit has a definite probability of it being selected in the sample, is known as "Probability sampling".

Q.23 A sampling procedure in which there is no probability of selection attached with the sampling units, is known as " non-probabilistic sampling.

Q.24 The enumeration of population through sampling method was frirst time proposed by "Laplace in 1783".

Q.25 Randomization gives a technique for valid and representative selection of sample from a parent population.

Q.26 The ratio of sample size n to population size N *i* e $\frac{n}{N}$ is known as "inclusion probability" of a unit in the sample.

Q.27 The selection probability of a unit from the population of size N remains contant and is equal to $\frac{1}{N}$.

1.3 Fill in the Blanks

Q.1 Sampling error occurs only in ___________ Survey.

Q.2 Schedule is a list of questions carried and filled in by an ________

Q.3 The bias of an estimator $t_n(x)$ is defined as ___________

Q.4 Collecting information on a small part of population, is known as ___________

Q.5 Non - sampling errors occur in ___________ and in _______ .

Q.6 A true sample should posses ___________ of the population under study.

Q.7 The sample proportaion proportion "p" is an _____ estimator of population proportion "P".

Q.8 Any population constent say 'μ' is also known as _______

Q.9 The difference between population value (parameter) and its estimated value arised due to sampling process adopted, is known as ___________

Q.10 If all the units of the population are enumerated then it is called ______

Q.11 The square of S.E. of an estimator is known as ___________

Q.12 The calculated value of an estimator from sample values is called as an ___________ of the population value.

Q.13 The sample value $s^2 = \frac{1}{(n-1)} \sum_{i}^{n} (x_i - \bar{x}_x)^2$ is known as _________

Q.14 The population value $S^2 = \frac{1}{(N-1)} \sum_{\lambda=1}^{n} (X_i - \bar{X}_N)^2$ is known as _______

Q.15 The expression to work out the sample variance s^2 is written as ___________

Q.16 A complete and upto date list of all units of the population with their identification to known as ___________

Q.17 A method based on a principle to optain an estimate of a quantity with a given level of accuracy with minimum possible resources is known as principle of ___________

Q.18 A complete enumeration is ___________

Q.19 A function of all the sampling units of the population is known as ___________

Q.20 A survey in which only a fraction of the population is enumerated, is called ___________

Q.21 The ratio of sample size n to the population size N, is known as ___________

Q.22 Precision of an estimator is _______________ proportional to its sampling variance.

Q.23 The variance of an estimator t_n (x) is known as ___________

Q.24 The arithmatic mean of square of deviations of individual units (elementary units) fram their arithmatic mean is known as ___________ and is denoted by ___________

Q.25 The sampling variance of an unbiased estimator t_n (x) of the parameter μ can be defined as _______________

1.4 Multiple Choice Questions and Answers

Q.1 A function of all sampling units in the sample is known as-

(a) Parameter (b) Statistic

(c) Estimator (d) Estimate

Q.2 Random number table is used to

(a) To draw a random sample (b) To define a statistic

(c) To decide the sample size n (d) None of these

Q.3 Collection of all sampling units in the population is known as

(a) Sampling frame (b) Population

(c) Parameter (d) None of these

Q.4 Sample survey is -

(a) Less costly than census

(b) More scopic than complete enumeration

(c) Less scopic than complete enumeration

(d) Both (a) & (b)

Q.5 The ratio $f = \dfrac{n}{N}$ is known as

(a) Sampling fraction

(b) Inclusion probability of a unit in the sample

(c) Raising factor

(d) Both (a) & (b)

Q.6 Precision of an estimator is
(a) Proportional to its sampling variance
(b) Is versely proportional to its sampling variance
(c) Equal to its standard error
(d) None of these

Q.7 Sample survey may produce
(a) More accurate result than that of complete enumeration
(b) Less accurate result than that of complete enumeration
(c) More non–sampling error
(d) Less non-sampling error

Q.8 An estimator is a
(a) Statistic (b) Ramdom Variable
(c) Sample Value (d) Both (a) & (b)

Q.9 Totality of all sampling units is known as
(a) Sample (b) Population
(c) Sampling frame (d) Population total

Q.10 Sampling units may be
(a) Individuals (b) Group of individuals
(c) Patients (d) All the three

Q.11 A list of all sampling units with their identification is known as
(a) Sampling frame (b) Population
(c) Census (d) Parameter

Q.12 A function of all sampling units of the population is known as
(a) Population parameter (b) Population constant
(c) An estimator (d) Both (a) & (b)

Q.13 Sample statistic can be considered as
(a) An estimator (b) A romdom variable
(c) A test statistic (d) All the three as needed

Q.14 Enumeration of a small part of population is known as
(a) Census (b) Sample Surveys
(b) Sample (d) Sample Statistic

Q.15 One of the roles of sampling theory is -

(a) To provide an estimate of sampling variance
(b) To provide an estimate of M. S. E. of an estimator
(c) To provide an estimate of S. E. of an estimator
(d) All the three (a), (b) and (c)

Q.16 A sampling procedure which is governed by a sampling rule or is independent of sampler's discretion, is known as-

(a) Objective sampling
(b) Subjective sampling
(c) Purposive sampling
(d) All the three

Q.17 A sampling procedure in which every units has a preassigned definite probability of it being selected in the sample, is known as

(a) Purposive sampling
(b) Probability sampling
(c) Scientific sampling
(d) Subjective sampling

Q.18 A sampling procedure in which there is no probability of selection attached with the sampling units, is known as-

(a) Non-probabilitic sampling
(b) Judgement sampling
(c) Systematic sampling
(d) Optimistic sampling

Q.19 Enumeration of population by sampling method was first time proposed by-

(a) Lalace in 1783
(b) Fisher R. A. in 1783
(c) P.C. Mahalanobia in 1783
(d) Sukhatme P.V. in 1783

Q.20 The ratio of sample size n to population size N is known as

(a) Inclusion probability
(b) Sampling fraction
(c) Raising factor
(d) Both (a) and (b)

Q.21 The selection probability of a unit from the population of size N remains constant and is equal to

(a) $\left(1-\frac{1}{N}\right)$
(b) $\frac{1}{N}$
(c) $\frac{n}{N}$
(d) $\frac{N}{n}$

Q.22 Sampling error occurs in

(a) Complete enumeration only (b) Sample survey only
(c) (a) and (b) both (d) Random sampling

Q.23 Schedule is a list of questions carried and filled by an-

(a) Investigator (b) Superviser
(c) Field staff (d) Statistician

Q.24 The bias of an estimator $t_n(\underline{x})$ of parameter θ is defined as.

(a) Q -E $(t_n(\underset{\sim}{x}))$ (b) E $(t_n(\underset{\sim}{x}))$ - θ
(c) $\hat{\theta}-\theta$ (d) $\theta-\hat{\theta}$

Q.25 Non - Sampling error occurs in

(a) Sample survey (b) Complete enumeration
(c) Systematic sampling (d) Both in (a) and (b)

Q.26 A true sample must possess the properly of

(a) Representativeness of population.
(b) Unbiasedness of population.
(c) Biasedness of population
(d) Consistency of population

Q.27 The sample proportion "P" is an ___________ estimator of population proportaion "P"

(a) Biased (b) Consistent
(c) Unbiased (d) None of these

Q.28 Any population constent is also known as

(a) Statistic (b) Estimator
(c) Parameter (d) All the three

Q.29 The difference between the population parameter and its estimated value based on sample value to known as -

(a) Sampling error (b) Standard error
(c) Standard deviation (d) Biasness

Q.30 Enumeration of all the sampling units of the population is known as

(a) Sample surveys (b) Complete enumeration

(c) Census (d) (b) and (c)

Q.31 The square of S. E. of an estimator is known as

(a) Sampling variance (b) Population variance

(c) Population mean deviation (d) None of these

Q.32 The calculated value of an estimator from sample values is known as

(a) An estimate (b) Statistic

(c) Sample variance (d) None of these

Q.33 An expression to work out the sample variance s^2 is given by

(a) $s^2 = \frac{1}{n}\sum_{i}^{n}(y_i - \bar{y}_n)^2$ (b) $s^2 = \frac{1}{(n-1)}\sum_{i=1}^{n}(y_i - \bar{y}_n)^2$

(b) $s^2 = \frac{1}{N}\sum_{i}^{N}(y_i - \bar{y}_N)^2$ (d) $s^2 = \frac{1}{(N-1}\sum^{N}(y_i - \bar{y}_N)^2$

Q.34 A complete and up to date list of all units of population with their identification is known as-

(a) Population (b) Sampling frame

(c) Population parameter (d) Complete enumeration.

Q.35 A method to obtain in estimate of a quantity with a given level of accuracy with minimum possible resources is known as principle of -

(a) Least square (b) Optimization

(c) Regularity (d) Consistency

Q.36 Complele enumeration is ________________ sample survey.

(a) More efficient than (b) Less efficient than

(c) More costly than (d) Both (b) & (c)

1.5 Key Answers

1.5.1 (True/False)

1.	False	2.	False	3.	False
4.	True	5.	True	6.	False

7.	False	8.	True	9.	False
10.	True	11.	True	12.	True
13.	False	14.	False	15.	False
16.	False	17.	False	18.	False
19.	True	20.	True	21.	True
22.	True	23.	True	24.	True
25.	True	26.	True	27.	True

1.5.2 Fill in the Blanks

1. Sample
2. Investigator
3. E $(t_n(\underline{x}))$ - Parametric value
4. Sample Surveys
5. Sample survey, Complete enumeration
6. Property of representativeness.
7. Unbiased
8. Population Parameter
9. Sampling error
10. Complete enumeration or census
11. Sampling variance
12. Estimate
13. Sample mean square or sample variance
14. Population mean square
15. $s^2 = \frac{1}{(n-1)}\sum_{i=1}^{n}(y_i - \bar{y}_n)^2$
16. Sampling frame
17. Optimization
18. Costly, Time consuming, Less efficient, of less use, of less scope
19. Population parameter
20. Sample Surveys
21. Inclusion probability of a unit in the sample.
22. Inversely
23. Sampling variance

24. Population Variance s2

25. $V\ (t_n(x)) = E(t_n(\underset{\sim}{x}) - \mu)^2$

1.5.3 Multiple Choice Questions and Answers

1. (b) Statistic
2. (a) To draw a random sample
3. (b) Population
4. (d) Both (a) & (b)
5. (d) Both (a) & (b)
6. (b) Inversely proportional to its sampling variance
7. (d) Both (a) and (d)
8. (d) Both (a) & (b)
9. (b) Population
10. (d) All the three
11. (a) Sampling frame
12. (d) Both (a) & (b)
13. (d) All the three as needed
14. (b) Sample Surveys
15. (d) All the three (a), (b) & (c)
16. (a) Objective Sampling
17. (b) Probability Sampling
18. (a) Non-probalistic sampling
19. (a) Laplace in 1783
20. (d) Both (a) & (b)
21. (d) $\frac{1}{N}$, Both (a) & (b)
22. (b) $\frac{1}{N}$
23. (b) Sample Surveys only
24. (b) $E\ (t_n\ (\underset{\sim}{x})) - Q$

25. (d) Both (a) & (b)
26. (a) Representiveness of population
27. (c) Unbiased
28. (c) Parameter
29. (a) Sampling error
30. (d) Both (b) & (c)
31. (a) Sampling variance
32. (a) An estimate
33. (b) $s^2 = \frac{1}{(n-1)}\sum_{i}^{n}(y_i - \overline{y}_n)^2$
34. (b) Sampling frame
35. (b) Optimization
36. (d) Both (b) & (c)

2

Sample Random Sampling (S.R.S.W.O.R. & S.R.S.W.R.)

2.1 True / False

Q.1 SRSWOR scheme is better (more efficient) them SRSWR scheme.

Q.2 The sampling variance of sample mean under SRSWOR scheme is more than that in SRSWR scheme.

Q.3 The sample variance s^2 is an unbiased estimate of population M.S. (S^2) in case of SRSWOR scheme.

Q.4 The variance of sample mean $\overline{y}_n$ decreases with an increase of sample size n in SRSWOR as well as in SRSWR schemes.

Q.5 The sampling procedure SRSWR is not better than SRSWOR scheme.

Q.6 s^2 is an unbiased estimate of s^2 in SRSWOR scheme.

Q.7 The sampling variance of sample mean $\overline{y}_n$ decreases with an increase in sample size n.

Q.8 In SRSWR scheme, for any subsequent draw, the population size is decreased by one.

Q.9 An estimator is said to be positively biased to the parameter if its expected value is more than the parameter.

Q.10 The sample mean square (s^2) is biased estimator of population mean square (S^2) in case of SRSWR.

Q.11 SE. ($\overline{y}_n$) SRSWOR > S. E. ($\overline{y}_n$) SRSWR.

Q.12 Remainder method of sample selection is better than that of random number method.

Q.13 The sample mean $\overline{y}_n$ is not an unbiased estimate of population mean $\overline{y}_n$ under SRSWOR scheme.

Q.14 The sample variance (s^2) is an unbiased estimate of population variance s^2 in SRSWOR scheme.

Q.15 An estimate of sampling variance $(\overline{y}_n)_{SRSWOR}$ is expressed as V $(\overline{y}_n)_{SRSWOR} = \frac{(N-1)}{nN} s^2$

Q.16 S. E. $(\overline{y}_n)_{SRSWOR} = \sqrt{\frac{N-1}{nN} s^2}$

Q.17 The mathematical relation among M. S. E. $(\overline{y}_R)$, V $(\overline{y}_R)$ and Bias $(\overline{y}_R)$ exists as

M. S. E. $(\overline{y}_R)$ = V $(\overline{y}_R)$ + (Bias $(\overline{y}_R)$)^2

Q.18 Sampling scheme SRSWOR is less efficient tham SRSWR Scheme.

Q.19 The simplest and most common method of sampline techniques is the simple random sampling.

Q.20 In usual notations and under SRSWOR scheme, the number of all possible samples is equal to ${}^{N}C_{n}$.

2.2 Fill in the Blanks

Q.1 The sample proportion 'p' is an ________________ estimate of population proportion 'P' under S.R.S.W.O.R. scheme.

Q.2 The formula to workout V$(\overline{y}_n)_{SRSWOR}$ in usual notations is written as ________________

Q.3 An unbiased estimate of population variance s^2 in case of SRSWR is worked out from the expression written as ________

Q.4 A consistent estimate of S.E. $(\overline{y}_n)_{SRSWOR}$ can be obtained from an expression ________________________

Q.5 An unbiased estimate of V$(\overline{y}_n)_{SRSWOR}$ can be obtained from an expression ____________________________

Q.6 S. E. $(\overline{y}_n)_{SRSWOR}$

Q.7 For a population of size N = 5 units given as y : 11, 2, 7, 15 and 9, the value of population mean square (S^2) is equal to ________________

Q.8 The sample statistic (s^2) is an unbiased estimator of population variance s^2 in case of sampling scheme known as ______________

Q.9 There is a sample of n=5 size and values as y: 11, 2, 7, 15 and 9, then the value of sample variance (s^2) is equal to ___________

Q.10 An unbiased estimate of population mean $\bar{Y}_N$ in case of SRSWOR is given by ______________

Q.11 $V(\bar{y}_n)_{SRSWOR}$ is ______________ then $V(\bar{y}_n)_{SRSWR}$

Q.12 In usual notations n, and N under SRSWR, the number of all possible samples is equal to ______________

Q.13 The standared deviation worked out from all possibles sample means $\bar{y}_n$ of sizn n drawn from population of size N is known as ____________ of the estimator $\bar{y}_n$

Q.14 If the observations recorded on 6 units of a sample are as y : 3, 4, 5, 7, 8, 9 then the sample variance (s^2) is equal to ____________

Q.15 If all values of a sample, are same then the value of sample variance (s^2) will be equal to ______________

Q.16 The ratio $\frac{(N-n)}{N}$ is known as ___________

Q.17 The parameter $S^2 = \frac{1}{(N-1)}\sum_{\ell=1}^{N}(y_i - \bar{Y}_N)^2$ is known as ____________

Q.18 The bias of an estimator $t_n(x)$ is defined as ____________ to estimate the population parameter θ.

Q.19 A consistant estimate of S. E. $(\bar{y}_n)_{SRSWOR}$ can be obtained from an expression ____________

Q.20 An unbiased estimator of V $(\bar{y}_n)_{SRSWOR}$ can be written as _________

Q.21 An unbiased estimate of s^2 in case of SRSWR can be obtained from an expression written as ____________

Q.22 An estimator $t = N\bar{y}_n$ under SRSWOR gives an ___________ estimate of ____________

Q.23 Sample values are as y : 2, 4, 6, 8, 10. Then an unbiased estimate of population mean is equal to ___________

Q.24 The sample values y : 10, 8, 6, 4, 2, gives the value fo sample variance (s^2) equal to ____________

Q.25 From given statistics as $S^2 = 20$, n=4, N= 10, the $V(\overline{y}_n)$ SRSWOR is equal to ____________

Q.26 Out of two sampling schemes SRSWOR and SRSWR, the scheme which always gives a better representative sample is the ________

2.3 Multiple Choice Questions and Answers

Q.1 Sampling variance is the variance of

(a) An estimator (b) Individual units of population
(c) Sample statistic (d) None of these

Q.2 Sample variance s^2 under SRSWOR is an unbiased estimate of.

(a) Population MS (S^2),
(b) Population Variance (s^2)
(c) Sampling Variance of sample mean
(d) None of these

Q.3 Sample variance s^2 is an unbiased estimate under SRSWR scheme of.

(a) Population variance s^2 (b) Population M. S (S^2)
(c) Variance of sample mean (d) None of these

Q.4 The variance of sample mean $(\overline{y}_n)$ in case of SRSWOR decreases with -

(a) An increase in sample size (b) Decrease in sample size
(c) Unaffected with sample size (d) None of these

Q.5 In SRSWR scheme, for any subsequent draw, the population size-

(a) Is decreased by one (b) Is increased by one
(c) Remains constant (d) None of these

Q.6 If the expected value of an estimator is greater than the value of a parameter then the estimator is said to be-

(a) Negatively biased (b) Positively biased.
(c) Unbiased (d) None of these

Q.7 In case of sampling variance of sample mean $\overline{y}_n$, the following relation exists.

(a) $V\ (\overline{y}_n)_{SRSWOR} > V\ (\overline{y}_n)_{SRSWR}$

(b) $V\ (\overline{y}_n)_{SRSWOR} < V\ (\overline{y}_n)_{SRSWR}$

(c) $V\ (\overline{y}_n)_{SRSWOR}\ ^3\ V\ (\overline{y}_n)_{SRSWR}$

(d) None of these

Q.8 An unbiased estimate of $V\ (\overline{y}_n)_{SRSWOR}$ can be expressed as

(a) $\hat{V}(\overline{y}_n) = \dfrac{N-1}{nN} s^2$ (b) $\hat{V}(\overline{y}_n) = \dfrac{N-n}{nN} s^2$

(c) $\hat{V}(\overline{y}_n) = \dfrac{N-n}{nN} s^2$ (d) $\hat{V}(\overline{y}_n) = \dfrac{N-1}{nN} s^2$

Q.9 A consistent estimate of $S.\ E.\ (\overline{y}_n)_{SRSWOR}$ can be written as

(a) $\sqrt{\dfrac{N-1}{nN}s^2}$ (b) $\sqrt{\dfrac{N-n}{nN}s^2}$

(b) $\sqrt{\dfrac{N-n}{nN}s^2}$ (d) $\sqrt{\dfrac{N-1}{nN}s^2}$

Q.10 For a biased estimator t, the following relation exists.

(a) M S E. (t) = V (t) + Bias (t)

(b) M. S E (t) = V (t) + $(\text{Bias}\ (t))^2$

(c) V (t) = M. S E (t) + Bias (t)

(d) V (t) = M.SE (t) + $(\text{Bias}\ (t))^2$

Q.11 In SRSWOR, the number of all possible samples is equal to

(a) ${}^{N}C_{n}$ (b) N^n

(c) $\dfrac{N}{n}$ (d) $\dfrac{N}{n}$

Q.12 The sampling fraction is defined as

(a) $\frac{N}{n}$ (b) $\frac{n}{N}$

(c) $\overset{N}{C}_n$ (d) $\frac{N-n}{N}$

Q.13 Finite population correction is defined as

(a) $\frac{N-1}{N}$ (b) $\frac{N-n}{n}$

(c) $\frac{N-n}{N}$ (d) $\frac{N-1}{n}$

2.4 Key Answers

2.1.1 True / False

1.	True	2.	False	3.	True
4.	True	5.	True	6.	False
7.	True	8.	False	9.	True
10.	True	11.	False	12.	True
13.	False	14.	False	15.	False
16.	False	17.	True	18.	False
19.	True	20.	True		

2.4.2 Fill In The Blank

1. Unbiased

2. $\frac{N-n}{nN}S^2$

3. $s^2 = \frac{1}{n-1}\sum_{i}^{n}(y_i - \bar{y}_n)^2$

4. $\sqrt{\frac{N-n}{n-N}s^2}$

5. $\frac{N-n}{nN}s^2$

6. $\sqrt{\frac{N-n}{nN}s^2}$

7. 23.2
8. SRSWR
9. 23.2
10. $\overline{y}_{\mathrm{n}} = \frac{1}{\mathrm{n}} \sum_{i}^{n} y_i$
11. Loss
12. N^n
13. Standard error
14. 4.666
15. Zero
16. Finite population correction
17. Population mean square
18. E $(t_n(\underset{\sim}{x})) - Q$
19. $\sqrt{\frac{N-n}{nN}} s^2$
20. $\frac{N-n}{nN} s^2$
21. $s^2 = \frac{1}{(n-1)} \sum^{n} (y_i - \overline{y}_n)^2$
22. Unbiased, Population total Y,
23. 6.0
24. 12.00
25. 3.0
26. SRSWOR

2.4.3 Multiple Choice Questions and Answers

1.	a	2.	a	3.	a
4.	a	5.	c	6.	b
7.	b	8.	b	9.	b
10.	b	11.	a	12.	b
13.	c				

3

Stratified Random Sampling

3.1 True / False

Q.1 If the sampling units of the population are of heterogeneous nature then stratified random sampling scheme is suitable for sample survey to be conducted.

Q.2 In stratatified random sampling scheme, the strata are homogeneous within themselves but heterogeneous between themselves.

Q.3 In stratified random sampling, we select samples of different sizes from different strata independently.

Q.4 Division of heterogeneous population into certain number of homogeneous groups is known as statification.

Q.5 In each stratum, we use SRSWOR to select independent random samples of different size depending upon stratum size. This sampling procedure adopted is known as "Stratified Random Sampling".

Q.6 In stratified random sampling adopted, the different samples drawn from different strata are known as "Stratified Random Samples".

Q.7 In adopting stratified random sampling, the strata should be overlaping and should not comprise together the whole population.

Q.8 In many practical situations, the administrative convenience may be considered as the basis for stratification.

Q.9 Stratification w. r. to natural characteristics, helps in improving the sampling design.

Q.10 Stratification is particularly more effective when there are extreme values in the population which can be segregated into separate strata.

Q.11 Stratification does not enable the statistician to use different sampling designs in different strata.

Q.12 Stratification does not ensure adequate representation to various parts of population in the sample.

Q.13 Stratification obtains a gain in the precision of an estimate of population characteristic under estimation.

3.2 Notations

Let the population be divided into k strata such that the size of its stratum is N_i (i = 1, 2, ...k). From each of these strata, independent random samples of sizes n_i (i = 1, 2, ...k) are drawn using SRSWOR since strata are non-overlaping, we have

$$N = \sum_{i=1}^{k} N_i \quad = \text{Population size}$$

$$n = \sum_{i=1}^{k} n_i \quad = \text{Total sample size}$$

$$\overline{Y}_N = \sum_{i=1}^{k} W_i \overline{Y}_{Ni} = \sum_{i}^{k} \frac{N_i}{N} \overline{Y}_{Ni} = \text{Population mean}$$

y_{ij} : The value of jth unit in the ith stratum.

Q.14 In usual notations, the population mean in stratified random sampling is defined as $\overline{Y}_N = \sum_{i=1}^{k} W_i \overline{Y}_{Ni}$

Q.15 The stratum mean $\overline{Y}_{Ni}$ can be estimated unbiasedly by the sample mean $\overline{y}_{ni} = \frac{1}{n_i} \sum_{i=1}^{ni} y_{ij}$

Q.16 The population mean $\overline{Y}_N$ can be estimated unbiasedly by the weighted sample maen $\overline{y}_W$ defined as $\overline{y}_W = \sum_{i=1}^{k} w_i \overline{y}_{ni}$

Q.17 An unbiased estimate of population total can be proposed as $\hat{Y} = \sum_{i=1}^{k} N_i \overline{y}_{ni}$

Q.18 The sampling variance of the weighted sample mean $\overline{y}_w$ can be written as

$$V(\overline{y}_w) = \sum_{i=1}^{k} W_i^2 \frac{(N_i - n_i)}{n_i N_i} S_i^2$$

Q.19 The sampling variance of the sample mean $\overline{y}_{ni}$ drawn from its stratum is written as

$$V\ (\overline{y}_{ni}) = \frac{(N_i - n_i)}{n_i\, n_i} \times S_i^2$$

Q.20 The criteria for allocating the sample sizes is to minimise the budget for a given precision or maximise the precision for a fixed budget.

Q.21 In case of equal allocation, the sample size ni (i = 1, 2,....,k) will be same for all strata and will be worked out as

$$n_i = \frac{n}{k},\ (i = 1, 2, \ldots\ldots, k)$$

Q.22 The proportional allocation procedure is very common in practice and n_i (i = 1, 2, ...,k) is worked out as

$$n_i = W_i\, n, \left(W_i = \frac{N_i}{N}\right)$$

Q.23 Neyman allocation (1934) is also called as minimum variance allocation and is presented as

$$n_i = n \frac{N_i S_i}{\sum_i^n N_i S_i} = n \frac{W_i S_i}{\sum_i^k W_i S_i}$$

Q.24 In case of optimum allocation, we have to take a larger sample if

(i) N_i is larger (ii) S_i is larger

(iii) C_i (cost of surveying per unit) is cheaper

Q.25 In case of proportional allocation method, $V(\overline{y}_w)$ is written as

$$V(\overline{y}_w)\ \text{prop} = \frac{N - n}{n\,N} \sum_{i=1}^{k} W_i S_i^2$$

Q.26 In case of proportional allocation procedure, estimate of $V(\overline{y}_w)$ is written as

$$\hat{V}(\overline{y}_w)\ \text{prop} = \left(\frac{N - n}{n\ N}\right) \sum_{i=1}^{k} W_i\, s_i^2,\quad E(s_i^2) = S_i^2$$

Q.26 In case of Neyman allocation procedure, $V(\overline{y}_w)$ is written as

$$V(\overline{y}_w)_{Ney} = \frac{1}{n}\left(\sum_i^k W_i \; s_i)\right)^2 - \frac{1}{N}\sum_i^k W_i \; s_i^2$$

Q.27 In case of Neyman allocation, an estimate of $V(\overline{y}_w)$ can be written as

$$\hat{V}(\overline{y}_w) = \frac{1}{n}\left(\sum_i^k W_i \, s_i\right)^2 - \sum_i^k W_i \, s_i^2$$

Q.28 There is not gain in precision with stratified random sampling adopted through proportional allocation as compared to SRS scheme if strata means differ considerably.

Q.29 The gain in precision due to applying Neymon allocation as compared to SRS scheme, will be considerably large for the two following differences being large.

(i) $\left(\overline{Y}_{wi} - \overline{Y}_{ni'}\right), i \neq i'$ (ii) $\left(Si - S_{i'}\right), \; i \neq i'$

Q.30 The gain in precision due to adopting Neyman allocation as compared to proportional allocation will be larger if the difference (S_i - S_i ') is larger for $i \neq i'$.

Q.31 Neyman allocation procedure can not be adopted if the values of (S_i, i = 1, 2, ... , k) are known in prior.

Q.32 Sometimes, it may happen that in case of adopting Neyman allocation, the optimum values of ni in any stratum may happen to be greater them N_i. In that situation, we take ni = Ni

3.3 Fill in the Blanks

Q.1 If the sampling units of the population are of heterogeneous nature then for conducting the sample survey, the suitable sampling scheme is ____________.

Q.2 While doing stratification in case of heterogeneous population, the strata must be ____________ between themselves but ____________ within themselves.

Q.3 In stratified random sampling, we select samples of ____________ sizes from different strata independently.

Q.4 The process of dividing heterogeneous population into certain number of homogeneous subpopulations, is known as ____________

Q.5 In adopting stratified random sampling, the strata should comprise together the __________

Q.6 In many practical situations, the administrative convenience may be considered as the basis for __________.

Q.7 Stratification w. r. to natural characteristic helps in improving the __________.

Q.8 Stratification enables the statistician to use the different __________ in different strata.

Q.9 Stratification obtains a gain in the __________ of an estimate of population characteristic under study.

Q.10 In usual notations, the population mean under stratified random sampling is defined as __________

Q.11 The stratum mean $\overline{Y}_{Ni}$ can be estimated unbiasedly by sample mean defined as __________

Q.12 The population mean $\overline{Y}_N$ can be estimated unbiasedly by weighted sample mean $\overline{y}_w$ defined as __________

Q.13 An unbiased estimate of population total Y can be proposed as __________

Q.14 Sampling variance of a weighted sample mean $\overline{y}_w$ in stratified random sampling scheme can be expressed as __________

Q.15 The sampling variance of a sample mean $\overline{y}_{ni}$ drawn from h^{th} stratum will be written as __________

Q.16 The criteria for allocating the sample size is to minimise the budget (cost) for a given __________ of the estimate.

Q.17 In case of equal allocation, the sample size n_i (i =1, 2, - k) will be the __________ and can be worked out from the expression n_i = __________.

Q.18 In proportional allocation method, the sample size ni can be obtained from the expression written as: ni = __________

Q.19 Neyman allocation method is also known as __________ method.

Q.20 In Neyman allocation procedure, the sample size ni can be obtained as ni = __________

Q.21 In case of optimum allocation, we have to take larger sample if Ni and Si both are __________

Q.22 In case of optimum allocation, we take larger sample if C_i, the cost of surveying per unit is __________

Q.23 In case of proportional allocation, the sampling variance of the weighted sample mean $\overline{y}_w$ is written as __________

Q.24 In case of Neyman allocation $V(\overline{y}_w)$ can be written as __________

Q.25 In case of Neyman allocation, an estimate of $V(\overline{y}_w)$ can be obtained from the expression given as $\hat{V}(\overline{y})_{Ney} =$ __________

Q.26 There will be gain in precision in stratified random sampling through proportional allocation as compared to S. R. S scheme if strata means are __________.

Q.27 The gain in precision due to applying Neyman allocation as compared to S. R. S. scheme will be considerably large if the differences $(\overline{Y}_{Ni} - \overline{Y}_{Ni'})$ and $(Si - S_{i'})$, $i \neq i'$ are considerably __________.

Q.28 The gain in precision due to adopting Neyman allocation will be larger as compared to proportional allocation if differences $(S_i - S_{i'})$, $i \neq i'$, will be __________.

Q.29 To adopt Neyman allocation procedure, one must have a prior knowledge of __________.

Q.30 In adopting Neyman allocation, if it happen that ni > Ni then we should take ni = __________.

3.4 Multiple Choice Questions and Answers

Q.1 If the sampling units of population are haterogeneous in nature, the appropriate sampling scheme adopted for sample survey is

(a) Cluster Sampling (b) Stratified Random Sampling

(c) Systematic Sampling (d) Cluster Sampling

Q.2 In stratified random sampling, the strata must be

(a) Homogeneous within themselve

(b) Homogeneous between themselves

(c) Heterogeneous between themselves
(d) Both (a) & (c)

Q.3 In stratified random sampling, the samples from different strata are selected-

(a) Systematically (b) Independently
(c) Inter dependently (d) In clustering

Q.4 Division of heterogeneous population into certain number of homogeneous subpopulations to conduct sample survey is known as

(a) Classification (b) Booking
(c) Stratification (d) Clustering

Q.5 In adopting stratified random sampling, the strata should be.

(a) Nonoverlaping (b) Overlaping
(c) Should comprise together the whole population (d) Both (a) and (c)

Q.6 Stratification enables the statistician to use different sampling designs in -

(a) In same stratum (b) Different strata
(c) Different sub-populations (d) Both (b) and (c)

Q.7 Stratification ensures adequate representation to various parts of population in -

(a) The samples (b) Strata
(c) Sub-population (d) Stratum

Q.8 Stratification gains in the -

(a) Precision of an estimate
(b) Sampling variance of an estimate
(c) S. E. of an estimate
(d) Unbiasedness of an estimate

Q.9 In usual notions, the population mean in stratified random sampling is defined as

(a) $\overline{Y}_N = \sum_{i=1}^{k} W_i \, \overline{y}_{ni}$ (b) $\overline{Y}_N = \sum_{i=1}^{k} W_i \, \overline{y}_{ni}$

(c) $\overline{Y}_N = \frac{\sum_{i}^{k} N_i \, \overline{y}_{N_i}}{N}$ (d) $\overline{Y}_N = \frac{\sum_{i}^{k} N_i \, \overline{y}_{ni}}{N}$

Q.10 The stratum mean $\overline{Y}_{Ni}$ ($i = 1, 2, ... k$) can be estimated by an estimator definded as

(a) $t = \frac{1}{n_i}\sum_{j=1}^{ni} y_{ij}$ (b) $t = \frac{1}{N_i}\sum_{j}^{Ni} y_{ij}$

(b) $t = \frac{1}{n}\sum_{j}^{k}\sum_{N=1}^{ni} y_{ij}$ (d) $\frac{1}{n}\sum_{j=1}^{ni} y_{ij}$

Q.11 The population mean $\overline{Y}_N$ can be estimated unbiasedly by an estimator defined as-

(a) $\overline{y}_w = \sum_{i=1}^{k} W_i \overline{y}_{Ni}$ (b) $\overline{y}_w = \sum_{i=1}^{k} W_i \overline{y}_{Ni}$

(c) $\overline{y}_w = \frac{1}{N}\sum_{i}^{K} N_i \overline{y}_{ni}$ (d) By both (b) and (c)

Q.12 The sampling variance of the estimator $\overline{y}_w$ can be worked out by the expression written as-

(a) $V(\overline{y}_w) = \sum_{i=1}^{k}\left(\frac{N_i - n_i}{n_i N_i}\right)s_i^2$

(b) $V(\overline{y}_w) = \sum_{i=1}^{k}\frac{N_i - n_i}{n_i N_i} W_i^2 s_i^2$

(c) $V(\overline{y}_w) = \sum_{i=1}^{k}\left(1 - \frac{n_i}{N_i}\right)\frac{W_i^2 s_i^2}{n_i}$

(d) Both (a) and (c)

Q.13 The sampling variance of the sample mean $\overline{y}_{ni}$ drawn from i^{th} stratum can be worked as from the expression-

(a) $V(\overline{y}_{ni}) = \frac{(N_i - n_i)}{nN} s_i^2$ (b) $V(\overline{y}_{ni}) = \frac{(N_i - n_i)}{n_i N_i} s_i^2$

(c) $V(\overline{y}_{ni}) = (1 - f_i)\frac{s_i^2}{n_i}$ (d) Both (b) and (c)

Q.14 The sampling variance of the estimator $\overline{y}_w$ can be estimated unbiasedly by an expression written as

(a) $\hat{V}(\overline{y}_w) = \sum_{i=1}^{k} \left(\frac{N_i - n_i}{n_i N_i} \right) s_i^2$

(b) $\hat{V}(\overline{y}_w) = \sum_{i}^{k} \left(\frac{N_i - n_i}{n_i N_i} \right) W_i s_i^2$

(c) $\hat{V}(\overline{y}_w) = \sum_{i=1}^{k} \left(1 - \frac{n_i}{N_i} \right) \frac{W_i^2 s_i^2}{n_i}$

(d) Both (b) and (c)

Q.15 The sampling variance of the sample mean $\overline{y}_{ni}$ can be estimated unbiasedly through the expression written as-

(a) $\hat{V}(\overline{y}_{ni}) = \frac{(N_i)}{N_i} s_i^2$ (b) $\hat{V}(\overline{y}_{ni}) = \frac{(N_i) s_i^2}{n N}$

(b) $\hat{V}(\overline{y}_{ni}) = (1 - f_i) \frac{s_i^2}{n_i}$ (d) Both (a) and (c)

Q.16 Criteria for allocating the sample sizes from different strata is to -

(a) To minimize budget for a given precision
(b) To maximize budget for a given precision
(c) To minimize precision for a given optimum budget
(d) To minimize budget for maximum precision

Q.17 In proportional allocation, the sample size n_i is equal to-

(a) $n_i = n.\ W_i$ (b) $n_i = Ni\ W_i$

(c) $n_i = n \frac{N_i}{N}$ (d) Both (a) and (c)

Q.18 In Neyman allocation, the sample sizes decided to be drawn from different strata are equal to-

(a) $n_i = n \frac{W_i S_i}{\sum W_i S_i}$ (b) $n \frac{W_i S_i}{\sum N_i S_i}$

(c) $n_i = n \frac{\sum^{k} N_i S_i}{\sum_{i=1} N_i S_i}$ (d) Both (a) and (c)

Q.19 In case of optimum allocation, we have to take a larger sample if

(a) N_i is larger and S_i is smaller

(b) N_i is smalller and S_i is larger

(c) N_i is larger and S_i is larger

(d) N_i is smaller and C_i is cheaper

Q.20 In case of applying proportional allocation method, the $V(\overline{y}_w)$ is equal to

(a) $V(\overline{y}_w)_{Prop} = \dfrac{N-n}{nN}\sum_{i}^{k} W_i s_i^2$

(b) $V(\overline{y}_w)_{Prop} = \dfrac{N-n}{nN}\sum_{i}^{k} W_i^2 s_i^2$

(c) $V(\overline{y}_w)_{Prop} = \dfrac{N-n}{nN}\sum_{i}^{k} (W_i s_i)^2$

(d) Both (b) and (c)

Q.21 In case of proportional allocation, $V(\overline{y}_w)$, can be estimated through.

(a) $\hat{V}(\overline{y}_w)_{Prop} = \dfrac{N-n}{nN}\sum^{k} (W_i s_i)^2$

(b) $\hat{V}(\overline{y}_w)_{Prop} = \dfrac{N-n}{nN}\sum_{i}^{k} W_i s_i^2$

(c) $\hat{V}(\overline{y}_w)_{Prop} = \dfrac{N-n}{nN}\sum_{i}^{k} \dfrac{N_i}{N} s_i^2$

(d) Both (b) and (c)

Q.22 In case of applying Neyman allocation, $V(\overline{y}_w)$ can be obtained through-

(a) $V(\overline{y}_w)_{Ney} = \dfrac{1}{n}\left(\sum_{k}^{k} W_i s_i\right)^2 - \dfrac{1}{N}\sum W_i s_i^2$

(b) $V(\overline{y}_w)_{Ney} = \dfrac{1}{n}\sum_{k}^{k} (W_i s_i)^2 - \dfrac{1}{N}\sum W_i s_i^2$

(c) $V(\overline{y}_w)_{Ney} = \frac{1}{n}\sum^{k}(W_i s_i)^2 - \frac{1}{N}\sum^{k} W_i^2 s_i^2$

(d) $V(\overline{y}_w)_{Ney} = \frac{1}{n}\left(\sum^{k} W_i s_i\right)^2 - \frac{1}{N}\sum W_i^2 s_i^2$

Q.23 In case of applying Neyman allocation, $V(\overline{y}_w)$ can be estimated through.

(a) $\hat{V}(\overline{y}_w)_{Ney} = \frac{1}{n}\sum_{i}^{k}(W_i s_i)^2 - \frac{1}{N}\sum^{k}(W_i s_i)^2$

(b) $\hat{V}(\overline{y}_w)_{Ney} = \frac{1}{n}\sum_{i}^{k}(W_i s_i)^2 - \frac{1}{N}\sum_{i}^{k} W_i s_i^2$

(c) $\hat{V}(\overline{y}_w)_{Ney} = \frac{1}{n}\sum_{i}^{k}(W_i s_i)^2 - \frac{1}{N}\sum_{i} W_i^2 s_i^2$

(d) $\hat{V}(\overline{y}_w)_{Ney} = \frac{1}{n}\sum^{k}(W_i s_i)^2 - \frac{1}{N}\sum_{i=1}^{k} W_i s_i^2$

Q.24 The gain in precision in estimating the population mean in case of applying Neyman allocation method as compared with applying SRSWOR will be high if -

(a) The difference $(\overline{Y}_{Ni} - \overline{Y}_{Ni'}), i \neq i'$ in larger

(b) The difference $(S_i - S_i)$, $i \neq i'$ is smaller

(c) Both the differences given in (a) and (b) are larger

(d) $(\overline{Y}_{Ni} - \overline{Y}_{Ni'})$ is larger and $(S_i - S_i)$ is smaller

Q.25 The gain in precision due to adopting Neyman allocation as compared to proportional allocation is higher if-

(a) The difference is higher $(\overline{Y}_{Ni} - \overline{y}_{Nj}), i \neq j,$ is highter

(b) The difference $(S_i - S_j), i \neq J,$ is higher

(c) The difference $(Y_{Ni} - Y_{Nj})$ *and* $(Si - Sj), i \neq j$ are smallar

(d) The difference $(S_i - S_j), i \neq J$ is smaller

3.5 Key Answers

3.5.1 True / False

1.	True	2.	True	3.	True
4.	True	5.	True	6.	True
7.	False	8.	True	9.	True
10.	True	11.	False	12.	False
13.	True	14.	True	15.	True
16.	True	17.	True	18.	True
19.	True	20.	True	21.	True
22.	True	23.	True	24.	True
25.	True	26.	True	27.	True
28.	False	29.	True	30.	True
31.	False	32.	True		

3.5.2 Fill in The Blanks

1 Stratified random sampling

2 Heterogeneous, homogeneous

3 Different

4 Stratification

5 Whole population

6 Stratification

7 Sampling design

8 Sampling designs

9 Precision (efficiency)

10 $\overline{y}_N = \sum_{h=1}^{k} W_i \overline{Y}_{Ni}$

11 $\overline{y}_{ni} = \frac{1}{n_i} \sum_{j}^{ni} y_{ij}$

12 $\overline{y}_w = \frac{k}{i=1} W_i \overline{y}_{ni}$

13 $\hat{y}_w = \sum_{i}^{k} N_i \overline{y}_{ni}$

14 $V(\overline{y}_w) = \sum_i^k W_i^2 \left(\frac{N_i - n_i}{n_i N_i} \right) s_i^2$

15 $V(\overline{y}_{ni}) = \frac{(N_i - n_i)}{n_i N_i} s_i^2$

16 Precision

17 Same, $n_i = \frac{n}{k}$

18 $n_i = n. W_i = n. \frac{N_i}{N}$

19 Optimum

20 $n_i = n. \frac{W_i s_i}{\sum_i^k W_i s_i}$

21 Larger

22 Cheaper

23 $V(\overline{y}_w)_{Prop} = \frac{N-n}{n N} \sum_{i=1}^k W_i s_i^2$

24 $V(\overline{y}_w)_{Ney} = \frac{1}{n} \left(\sum^k W_i s_i \right)^2 - \frac{1}{N} \sum W_i s_i^2$

25 $V(\overline{y}_w)_{Ney} = \frac{1}{n} \left(\sum_i^k W_i s_i \right)^2 - \frac{1}{N} \sum_i^k W_i s_i^2$

26 Differ considerably

27 Larger

28 Larger

29 Standard deviations S_i^2

30 $n_i = N_i$

3.5.3 Multiple Choice Questions and Answers

1. b	2. a	3. b
4. c	5. d	6. d
7. a	8. a	9. c

10.	a	11.	d	12.	d
13.	d	14.	d	15.	d
16.	a	17.	d	18.	d
19.	c	20.	a	21.	d
22.	a	23.	c	24.	c
25.	b				

4

Cluster Sampling

4.1 True/False

Q.1 In the theory of cluster sampling, if the population is divisible into smallest, distinct and identifiable units then this smallest unit is known as elementary unit (element) of the population.

Q.2 A group of elementary units in context of sample survey theory is known as cluster.

Q.3 When the clusters in the population are treated as sampling units then the sampling procedure adopted, is known as "cluster sampling".

Q.4 The main function of cluster sampling before the start of sample selection is to specify clusters or to divide the population into appropriate number of clusters.

Q.5 While preparing clusters, one must be careful that the number of elements in the cluster should be large and the number of clusters should be small.

Q.6 In context of adopting cluster sampling scheme, the complete enumeration is done in all the selected clusters.

Q.7 Cluster sampling is preferred not for convenience, easy work and economy point of view.

Q.8 For the valid use of cluster sampling, the necessary condition is that every unit of the population must belong to one and only cluster.

Q.9 While preparing the clusters in the population, no omission or-: duplication of elements in the cluster must take place to avoid the bias.

Q.10 Cluster sampling is generally more efficient than SRSWOR scheme due to general tendency of elements in a cluster being of similar nature.

Q.11 The efficiency of cluster sampling is likely to decrease with an increase in cluster size, Mi.

Q.12 The intraclass correlation coefficient ρ_w within the cluster is defined as

$$\rho_w = \frac{E(y_{ij} - \overline{y}_i)(y_{ij'} - \overline{y}_i)}{E(y_{ij} - \overline{y}_i)^2}$$ where yij is the value of jth element in the ith cluster and $\overline{y}_i$ is the mean of ith cluster.

Q.13 The limit of intraclass correlation coefficient ρ_w within a cluster is given as $0 \leq \rho_w \leq 1$

4.2 Notations for Equal Cluster Sizes

Suppose the population consists of N clusters, each of M elements and that a sample of n clusters is drawn by the method of S.R.S.W.O.R. Then we consider the following notations-

N : Number of clusters in the population

n : Number of clusters in the sample

M : Number of element in each cluster.

$\overline{y}_i = \frac{1}{M}\sum_{j}^{M} y_{ij}$ = The mean per element in the ith cluster.

$\overline{\overline{y}}_n = \frac{1}{n}\sum_{i=1}^{n} \overline{y}_i$ = Mean of cluster means in a sample of n clusters

$\overline{\overline{Y}}_N = \frac{1}{n}\sum_{i=1}^{n} \overline{y}_i$ = Mean of cluster mean in the population.

$\overline{y}_{..} = \frac{1}{N}\frac{1}{M}\sum_{i=1}^{N}\sum_{i=1}^{M} y_{ij}$ = Mean per dement in the population which is known as population mean.

$S_i^2 = \frac{1}{(M-1)}\sum_{j=1}^{M}(y_{ij} - y_i)^2$ = The mean square between the elements within i[th] (i=1, 2, N.) cluster.

$\overline{S}_w^2 = \frac{1}{N}\sum_{j=1}^{N} S_i^2$ = The mean square within cluster

$S_b^2 = \frac{1}{(N-1)}\sum_{j=1}^{N}(\overline{y}_i - \overline{\overline{y}}_N)^2$ = Mean square between cluster means in the population.

$$S^2 = \frac{1}{(NM-1)}\sum_{i}^{N}\sum_{j=1}^{M}(y_{ij}-\bar{y}_{..})^2$$ = Mean square between elements in the population.

$$\rho_w = \frac{E(y_{ij}-\bar{y}_{i.})(y_{ij'}-\bar{y}_{i.})}{E(y_{ij})-(\bar{y}_{i.})^2}$$

Q.14 In case of n clusters each of size M, drawn from N clusters using SRSWOR, an unbiased estimate of population mean $\bar{\bar{y}}_N$ is defined as

$$\bar{\bar{y}}_n = \frac{1}{n}\sum_{i=1}^{n} y_i$$

Q.15 The sampling variance of the sample mean $\bar{\bar{y}}_{ni}$ is defined as

$$V(\bar{\bar{y}}_n) = \left(\frac{N-n}{nN}\right) S_b^{\,2}$$

Where S_b^2 is the mean square between cluster means in the population.

Q.16 An expression to workout an unbiased estimate of $V(\bar{\bar{y}}_n)$ is written as

$$\hat{V}(\bar{\bar{y}}_n) = \left(\frac{N-n}{nN}\right) s_b^{\,2} \quad \text{Where } s_b^{\,2} = \frac{1}{(n-1)}\sum_{i=1}^{n}(\bar{y}_i-\bar{\bar{y}}_n)^2$$

Q.17 In terms of S^2 and $\bar{S}_w^{\,2}$, the $V(\bar{\bar{y}}_n)$ can be written as

$$V(\bar{\bar{y}}_n) = \frac{1}{nM(N-1)}[(NM-1)s^2 - N(M-1)\,\bar{S}_w^{\,2}].$$

Q.18 Cluster sampling is less efficient than SRS scheme because SRS provides a better cross section of population in the sample than that provided by cluster sampling.

Q.20 The sampling variance of the sample mean $\bar{\bar{y}}_n$ in term of intraclass correlation coefficient ρ_w is written as

$$V(\bar{\bar{y}}_n) = \frac{S^2}{nM}\{1+(M-1)\rho_w\} \text{ for sufficiently large N.}$$

Q.21 Cluster sampling and S R S becomes equivalent if M = 1.

Q.22 In cluster sampling, the intraclass correlation coefficient ρ_w is generally positive because in practice, the elements of the same cluster happen to be some how of similar nature.

Q.23 If sample of nM elements from the population are drawn using S. R. S scheme, then the relative efficiency of cluster sampling w. r. to S. R. S. is given by

$$\text{R. E. (cluster sampling / S R S)} = \frac{S^2}{M.S_b^{\ 2}}$$

Q.24 Efficiency of cluster sampling increases with the decrease of S_b^2 (ie M. S. between clusters).

Q.25 The relation among S^2, S_b^2 and $\overline{S}_w^{\ 2}$ exists as

$$M\,S_b^{\ 2} = \frac{1}{(N-1)}\,[(NM-1)S^2 - N\,(M-1)\,\overline{S}_w^{\ 2}\,]$$

4.3 Theory for Unequal Clusters

Let there be N clusters. Let the ith cluster consists of M_i elements (i = 1, 2, ... , N). and let $Mo = \sum_{i=1}^{N}$ Mi donates the number of elements in the population.

$$\bar{y}.. = \frac{1}{Mo}\sum_{i=1}^{N}\sum_{j}^{Mi} y_{ij} = \frac{1}{Mo}\sum_{i=1}^{N} M_i y_i$$

= Population mean or per unit population mean.

$$\bar{y}_i = \frac{1}{Mi}\sum_{j=1}^{Mi} y_{ij} = \text{Mean of ith cluster}$$

$$\bar{M} = \frac{Mo}{N} = \text{Average size of each cluster}$$

$$\bar{\bar{y}}_N = \frac{1}{N}\sum_{i}^{N} \bar{y}_i = \text{Pooled mean of the cluster means.}$$

Q.26 In general the pooled mean $\bar{\bar{Y}}_N \neq \bar{y}..$ = population mean.

Q.27 An estimator defined as $\bar{\bar{y}}_n = \frac{1}{n}\sum_{i}^{n} \bar{y}_i$ gives a biased estimate of population mean $\bar{\bar{y}}..$ in case of unequal cluster sizes.

Q.28 An estimator $\overline{y}n'$ defined as

$$\overline{y}_n' = \frac{1}{n}\sum_{i=1}^{n}\frac{M_i\, y_{i.}}{M}$$ gives an unbiased estimate of population mean $\overline{y}..$

Q.29 An estimator $\overline{y}_n''$ defined as (weighted mean of cluster means)

$$\overline{y}_n'' = \frac{\sum_{i}^{n} M_i\, y_{i.}}{\sum_{i=1}^{n} M_i}$$

Provides a consistent estimate of population mean $\overline{y}..$

Q.30 An expression to work out the amount of bias of the biased estimator $\overline{\overline{y}}_n$ is given as

$$\text{Bias } (\overline{\overline{y}}_n) = -\frac{(N-1)}{Mo} S_{M\overline{y}}$$

$$\text{Where } S_M\overline{y} = \frac{1}{(N-1)}(M_i - \overline{M})(\overline{y}_{i.} - \overline{\overline{y}}_N)$$

$$M(\overline{\overline{y}}_n) = \frac{(N-n)}{n\,N} S_b^{\,2} + \left(\frac{N-1}{Mo}\right)^2 S^2M\overline{y}$$

$$\text{and } S_b^2 = \frac{1}{(N-1)}\sum_{i=1}^{N}(y_{i.} - \overline{\overline{y}}_N)^2$$

Q.31 If Mi and $\overline{y}_{i.}$ are uncorrelated then bias $(\overline{\overline{y}}_n)$ will be equal to zero.

Q.32 If Mi and $\overline{y}_i.$ are uncorrelated then the estimater $\overline{\overline{y}}_n$ will provide an unbiased estimate of population mean $\overline{y}..$

Q.33 The amount of sampling variance $V(\overline{\overline{y}}_n)$ can be worked out from the expression

$$V(\overline{\overline{y}}_n) = \frac{N-n}{n\,N} S_b^{\,2}$$

Q.34 An expression to estimate the $V(\overline{\overline{y}}_n)$ is written as

$$\hat{V}(\bar{\bar{y}}_n) = \frac{N-n}{nN} s_b^{\,2}$$

Where $s_b^{\,2} = \frac{1}{(n-1)} \sum_{i=1} (\bar{y}_i - \bar{\bar{y}}n)^2$ and $E(s_b^2) = S_b^2$

Q.35 An expression to work out the $V(\bar{\bar{y}}n')$ is given as

$$V(\bar{\bar{y}}_n') = \frac{N-n}{nN} S''^{\,2}_b$$

Where $S''^{\,2}_b = \frac{1}{(N-1)} \sum_{i=1}^{n} \left(\frac{Mi}{\bar{\bar{M}}}\right)^2 (\bar{y}_{i.} - \bar{\bar{y}}..)^2$

Q.36 A consistent estimator of $V_1(\bar{\bar{y}}_n'')$ can be written as

$$\hat{V}(\bar{\bar{y}}_n') = \frac{N-n}{nN} s''^{\,2}_b$$

Where $s''^{\,2}_b = \frac{1}{(n-1)} \sum_{i=1}^{n} \left(\frac{M_i}{\bar{\bar{M}}}\right)^2 (\bar{y}_{i.} - \bar{\bar{y}}_n)^2$

Q.37 The sampling variance of an unbiased estimator $\bar{y}_n$ is written as

$$V(\bar{y}_n') = \frac{N-n}{(nN)} S'^{\,2}_b \quad \text{Where } S'^{\,2}_b = \frac{1}{(N-1)} \sum_{i}^{N} \left(\frac{Mi}{\bar{\bar{M}}}\bar{y}_{i.} - \bar{y}..\right)^2$$

Q.38 An expression to work out an unbiased estimate of sampling variance $V(\bar{y}_n')$ is written as

$$\hat{V}(\bar{y}_n') = \frac{N-n}{(nN)} s'^{\,2}_b$$

Where $s'^{\,2}_b = \frac{1}{(n-1)} \sum_{i}^{N} \left(\frac{Mi}{\bar{\bar{M}}}\bar{y}_{i.} - \bar{y}_n'\right)^2$, and $E\, s'^2_b = s'^2_b$

4.4 Fill in the Blanks

Q.1 In the Theory of cluster sampling, the smallest, distinct and identifiable unit is know as ______________

Q.2 A group of elementary units in context of sample survey theory is known as ______________

Q.3 The main function of cluster sampling before the start of sample selection is to specify clusters or to divide population into appropriate number of ______________

Q.4 In case of adopting cluster sampling for sample survey, the ______________ enumeration is done in all the selected clusters.

Q.5 For valid use of cluster sampling, the necessary condition is that every unit of the population must belong to ______________ cluster.

Q.6 While partitioning the population into clusters, no ommission or duplication of elements in the clusters must take place to avoide the ______________

Q.7 Cluster sampling is generally ______________ efficient than SRSWOR due to general tendency of elements of a cluster being of similar nature.

Q.8 The efficiency of cluster sampling is likely to decrease with an ______________ in cluster size M*i*.

Q.9 The limit of intraclass correlation coefficient ρ_w within a cluster is given as ______________

Q.10 In cluster sampling, the y_{ij} denotes the value of characteristic under study, of the *j* th ______________ in the ith ______________

Q.11 $\overline{y}_{i}.=\frac{1}{M}\sum_{j=1}^{M} y_{ij}$ denotes the ______________

Q.12 $\overline{\overline{y}}_{n}=\frac{1}{N}\sum_{j=1}^{n} y_{i.}$ denotes the ______________

Q.13 $\overline{y}.. \frac{1}{NM}\sum_{i=1}^{N}\sum_{j=1}^{M} y_{ij.}$ denotes the mean per element in the ______________

Q.14 $\overline{S}_{w}^{\,2}=\frac{1}{N}\sum_{i=1}^{N} S_i^2$ denotes mean square ______________ cluster in the ______________

Q.15 $\overline{S}_{b}^{\,2}=\frac{1}{(N-1)}\sum_{i=1}^{N}(\overline{y}_{i}-\overline{\overline{y}}_{N})^{2}$ denotes the mean square ______________ cluster in the ______________ .

Q.16 $S^2 = \frac{1}{(NM-1)}\sum_{i=1}^{N}\sum_{j=1}^{M}(y_{ij} - \bar{y}..)^2$ donotes mean square between ____________ in the ____________

Q.17 In usual notation of equal cluster sampling, the estimator $\bar{\bar{y}}_n = \frac{1}{n}\sum_{i}^{n}\bar{y}_i$ gives an ____________ estimate of population mean $\bar{y}_n$.

Q.18 The expression for $V(\bar{\bar{y}}_n)$ is written as $V(\bar{\bar{y}}_n) =$ ____________

Q.19 An unbiased estimate of $V(\bar{\bar{y}}_n)$ can be worked out from the expression written as.

$\hat{V}(\bar{\bar{y}}_n) =$ ____________

Q.20 To get minimum variance of $\bar{\bar{y}}_n$, the cluster must be as ____________ as possible.

Q.21 $V(\bar{\bar{y}}_n)$ in term of intraclass correlation coefficient ρ_w is written as ____________

$$V(\bar{\bar{y}}_n) = \frac{S^2}{nM}\{1 + \ldots\ldots\ldots\}$$ for sufficiently large N.

Q.22 Cluster sampling and S.R.S. becomes equivalent if M = ____________

Q.23 In cluster sampling we have R. E. (cluster sampling/SRS) $= \frac{S^2}{\ldots\ldots\ldots\ldots}$

Q.24 Efficiency of cluster sampling increases with ____________ in S_b^2.

Q.25 In case of unequal cluster size M*i*, the estimator $\bar{\bar{y}}_n = \frac{1}{n}\sum^{n}\bar{y}_i$ gives ____________ estimate of population mean $\bar{y}\ldots$

Q.26 An estimator $\bar{y}_n' = \frac{1}{n}\sum_{i}^{n}\left(\frac{Mi}{\bar{M}}\right)\bar{y}_i$ gives ____________ estimate of population mean $\bar{y}_N$.

Q.27 Bias $(\overline{\overline{y}}_n) = -$ ______________. $SM\overline{y}$

Q.28 $SM\overline{y} = \frac{1}{(N-1)} \sum_{i=1}^{N} (..............)(..............)$

Q.29 $M(\overline{\overline{y}}_n) = (..............)Sb^2 + (............)^2 S_{M\overline{y}}^2$

Q.30 If Mi and $\overline{y}i$ are uncorrelated then bias $(\overline{\overline{y}}n)$ will be equal to ______________

Q.31 If Mi and $\overline{y}_{i.}$ are uncorrelated then the estimator $\overline{\overline{y}}_n$ will provide an ______________ estimate of population mean.

Q.32 An expression to work out the $V(\overline{\overline{y}}_n)$ in case of unequal cluster sizes in written as

$V(\overline{\overline{y}}_n') = (\quad\quad) S_b^2$

Q.33 In case of unequal cluster sizes Mi (i =1, 2, ..., N), $S_b''^2$ is defined as

$$S_b''^2 = \frac{1}{(N-1)} \sum_{i=1}^{N} (.....)^2 (\overline{y}_{i.} - \overline{y}..)^2$$

Q.34 The expression for $V(\overline{y}n')$ of an unbiased estimator is written as $V(\overline{y}_n') = (......) S_b'^2$

Q.35 The expression for $S'b^2$ is given as

$$S_b'^2 = \frac{1}{(N-1)} \sum_{i}^{N} (\overline{y}_i - \overline{y}..)^2$$

Q.36 An unbiased estimate $\hat{V}(\overline{y}_n')$ can be written as.

$$\hat{V}(\overline{y}_n') = \frac{N-n}{(n\,N)} s_b^2$$

Where s_b^2 = ______________

4.5 Multiple Choice Questions and Answers

Q.1 In the theory of cluster sampling the smallest, distinct and identifiable units are known as

(a) Elementary units (b) Sampling units

(c) Elements (d) Both (a) and (c)

Q.2 A group of elementary units in context of sample survey theory is known as.

(a) Cluster (b) Sampling unit
(c) Sample (d) None of the three

Q.3 The main feature of the cluster sampling before the start of sample selection is-

(a) To specify the clusters
(b) To prepare the clusters
(c) To divide the population into appropriate number of clusters
(d) All the three (a) (b) & (c) are same

Q.4 In case of adopting cluster sampling, for sample surveys, the enumeration work is done as-

(a) All the elementary units of the selected cluster are enumerated.
(b) Appropriate fractions of each selected clusters are enumerated.
(c) Complete enumeration of each selected cluster is done.
(d) Either (a) or (c)

Q.5 For valid use of cluster sampling, the necessary condition is that each elementary unit of the population must belong to-

(a) Only one cluster, (b) Every cluster
(c) Only to selected clusters (d) Cannot be said with surity

Q.6 While preparing the clusters, no omission or duplication of elements in the clusters must take place to avaid the-

(a) Sampling error (b) Bias
(c) Standard error (d) None of these

Q.7 Efficiency of cluster sampling as compared with S.R.S.W.O.R. is

(a) More (b) Less
(c) Equivalent (d) None

Q.8 Efficiency of cluster sampling decreases as the cluster size M_i ____________

(a) Decreases
(b) Increases

(c) Becomes unequal

(d) Nothing can be said with surity

Q.9 Intraclass correlation coefficient ρ_w within a cluster lies in an interval given as

(a) $0 \leq \rho_w \leq 1$ (b) $0 \leq \rho_w \leq 2$

(c) $-1 \leq 1 \leq 1$ (d) $-1 \leq \rho_w \leq 0$

Q.10 In cluster sampling $\overline{y}_i = \frac{1}{M}\sum_{j=1}^{M} y_{ij}$ denotes the -

(a) Mean of ith cluster

(b) Sample mean of ith cluster.

(c) Sample mean drawn from the population

(d) None of these

Q.11 $\overline{y}_{..} = \frac{1}{NM}\sum_{ij}^{NM} yij$ in cluster sampling denotes the

(a) Mean per element in the population

(b) Mean of cluster means in the population

(c) Population means

(d) All the three defined in (a), (b) and (c)

Q.12 $\overline{S}_w^{\ 2} = \frac{1}{N}\sum_{i}^{N} S_i^2$ is known as

(a) Mean square within cluster

(b) Mean suqare between cluster

(c) Population mean square

(d) None of the three

Q.13 $S_b^2 = \frac{1}{N-1}\sum_{i=1}^{N}(y_i - \overline{\overline{y}}_N)^2$ denotes the

(a) Mean square between clusters in the population

(b) Mean square within clusters in the population

(c) Mean square of the population

(d) Mean square between clusters in the samples.

Q.14 The estimator $\overline{\overline{y}}_n = \frac{1}{n}\sum_{i}^{n} \overline{y}_i$ is the

(a) Unbiased estimator of population mean
(b) Biased estimator of population mean
(c) Consistent estimator of population mean
(d) None of the above three

Q.15 The sampling variance of the sample mean $\overline{y}_n$ is defined as-

(a) $V(\overline{\overline{y}}_n) = \frac{N-n}{nN} S_b^2$ (b) $V(\overline{\overline{y}}_n) = \frac{N-n}{nN} \overline{S}_w^2$

(c) $V(\overline{\overline{y}}_n) = \frac{n-N}{nN} S_b^2$ (d) $V(\overline{\overline{y}}_n) = \frac{N-n}{nN} S^2$

Q.16 An unbiased estimate of $V(\overline{\overline{y}}_n)$ can be obtained from

(a) $\hat{V}(\overline{\overline{y}}_n) = \frac{N-n}{nN} \overline{s}_w^2$ (b) $\hat{V}(\overline{\overline{y}}_n) = \frac{N-n}{nN} s_b^2$

(c) $\hat{V}(\overline{\overline{y}}_n) = \frac{n-N}{nN} s_b^2$ (d) $\hat{V}(\overline{\overline{y}}_n) = \frac{n-N}{nN} \overline{s}_w^2$

Q.17 To get the minimum variance of $\overline{\overline{y}}_n$, the clusters must be.

(a) As homogeneous as possible within themselves.
(b) As heterogeneous as possible within themselves.
(c) As homogeneous as possible between themselves.
(d) As heterogeneous as possible between themselves.

Q.18 The $V(\overline{\overline{y}}_n)$ in terms of ρ_w is written as for sufficiently large N

(a) $V(\overline{\overline{y}}_n) = \frac{S^2}{nM}\{1+(M-1)\rho_w\}$

(b) $V(\overline{\overline{y}}_n) = \frac{S_b^2}{nM}\{1+(M-1)\rho_w\}$

(c) $V(\overline{\overline{y}}_n) = \frac{S^2}{nN}\{1+(N-1)\rho_w\}$

(d) $V(\overline{\overline{y}}_n) = \frac{S^2}{nN}\{1+(M-1)\rho_w\}$

Q.19 Cluster sampling and simple random sampling becomes equivalent if

(a) M = 1 (b) M = 2

(c) n = N (d) None of these

Q.20 In cluster sampling, generally ρ_w is found to be-

(a) Positive (b) Negative

(c) $0 \le \rho w \le 1$ (d) Both (b) and (c)

Q.21 If sample of n M elements are drawn from population using S. R. S. scheme then. R. E (cluster sampling/ S. R. S.) is given by

(a) $\dfrac{S^2}{M\,\overline{S}_w^{\,2}}$ (b) $\dfrac{S^2}{M\,S_b}$

(c) $\dfrac{S^2}{M\,S_b^{\,2}}$ (d) $\dfrac{S_b^{\,2}}{M\,S^2}$

Q.22 Efficiency of cluster sampling increases if -

(a) $S_b^{\,2}$ decreases (b) $S_b^{\,2}$ increases

(c) $\overline{S}_w^{\,2}$ increases (d) Both (a) and (c)

Q.23 The relation among S^2, $\overline{S}_b^{\,2}$ and $\overline{S}_w^{\,2}$ exists as

(a) $M\,S_b^{\,2} = \dfrac{1}{(N-1)}[((NM-1)S^2 - N(M-1)\overline{S}^{\,2}{}_w]$

(b) $M\,S_b^{\,2} = \dfrac{1}{(N-1)}[(NM-1)S^2 - M(N-1)\overline{S}^{\,2}{}_w]$

(c) $M\,S_b^{\,2} = \dfrac{1}{(N-1)}[M(N-1)S^2 - N(M-1)\overline{S}^{\,2}{}_w]$

(d) $M\,S_b^{\,2} = \dfrac{1}{(N-1)}[N(M-1)S^2 - N(M-1)\overline{S}^{\,2}{}_w]$

Q.24 In case of unequal cluster size (ie mi), an unbiased estimator of population mean $\overline{y}..$ is defined as

(a) $\overline{y}_n' = \sum_i^n Mi\,\overline{y}_i / \sum_{i=1}^n Mi$ (b) $\overline{y}_n' = \dfrac{1}{n}\sum_i^n \left(\dfrac{Mi}{\overline{M}}\right)\overline{y}_i$

(c) $\overline{y}_n' = \dfrac{1}{n}\sum_i^n Mi\,\overline{y}_i$ (d) $\overline{y}_n' = \dfrac{1}{n}\sum_i^n \left(\dfrac{\overline{M}}{Mi}\right)\overline{y}_i$

Q.25 In case of unequal cluster sizes (Mi), a consistent estimater of population mean $\overline{y}..$ is defined as

(a) $\overline{y}_n'' = \frac{1}{n}\sum_{i=1}^{n}\left(\frac{Mi}{M}\right)\overline{y}_i$

(b) $\overline{y}_n'' = \frac{\sum_{i}^{N} Mi\, y_i}{\sum_{i}^{N} M_i}$

(c) $\overline{y}_n'' = \sum_{i}^{n}\left(\frac{Mi}{\overline{M}}\right)\overline{y}_i$

(d) $\overline{y}_n'' = \frac{1}{n}\sum_{i}^{n}\left(\frac{\overline{M}}{Mi}\right)\overline{y}_i$

Q.26 An expression to workout the bias of the biased estimator $\overline{\overline{y}}_n$ in case of unequal sizes, is given as

(a) Bias $(\overline{\overline{y}}_n) = \left(\frac{N-1}{M\,o}\right)S_{m\overline{y}}$

(b) $-\left(\frac{N-1}{M\,o}\right)S_{M\overline{y}}$

(c) $\left(\frac{N-1}{M\,o}\right)S_{M\overline{y}}$

(d) $-\left(\frac{M-1}{M\,o}\right)S_{M\overline{y}}$

Q.27 The covariance term $S_{M\overline{y}}$ between $(Mi, \overline{y}_i)$ is defined as

(a) $S_{M\overline{y}} = \left(\frac{1}{n-1}\right)\sum_{i}^{n}(Mi-\overline{M})(y_i-\overline{\overline{y}}_N)$

(b) $S_{M\overline{y}} = \left(\frac{1}{N-1}\right)\sum_{i}^{N}(Mi-\overline{M})(y_i-\overline{\overline{y}}_n)$

(c) $S_{M\overline{y}} = \left(\frac{1}{N-1}\right)\sum_{i}^{N}(Mi-\overline{M})(y_i-\overline{\overline{y}}_N)$

(d) $S_{M\overline{y}} = \left(\frac{1}{n-1}\right)\sum(Mi-\overline{M})(\overline{y}_i-\overline{\overline{y}}_n)$

Q.28 The M. S. E. $(\overline{\overline{y}}_n)$ will be obtained from-

(a) $M(\overline{\overline{y}}_n) = \frac{N-n}{n\,N}S_b^{\,2} + \left(\frac{\overline{M}}{M\,o}\right)^2 S^2{}_{M\overline{y}}$

(b) $M(\overline{\overline{y}}_n) = \frac{N-n}{n\,N}S_b^{\,2} + \left(\frac{N-1}{M\,o}\right)^2 S^2{}_{M\overline{y}}$

(c) $M(\overline{\overline{y}}_n) = \frac{N-n}{nN}\overline{S}_w^{\ 2} + \left(\frac{N-1}{Mo}\right)^2 S^2_{M\bar{y}}$

(d) $M(\overline{\overline{y}}_n) = \frac{N-n}{nN}\overline{S}_w^{\ 2} + \left(\frac{Mo}{\overline{M}}\right)^2 S^2_{M\bar{y}}$

Q.29 If Mi and $\overline{y}_i$ are uncorrelated then bias $(\overline{y}_n)$ will be

(a) Maximum (b) Minimum

(c) Zero (d) None

Q.30 If Mi and $\overline{y}_i$ are uncorrelated then the estimate of population mean obtained from an estimator $\overline{\overline{y}}_n$ will be-

(a) Biased (b) Unbiased

(c) Consistent (d) None

Q.31 The sampling variance $V(\overline{\overline{y}}_n)$ in case of unequal cluster size is written as.

(a) $V(\overline{\overline{y}}_n) = \frac{N-n}{nN}\overline{S}_w^{\ 2}$ (b) $V(\overline{\overline{y}}_n) = \frac{N-n}{nN}\overline{S}_b^{\ 2}$

(c) $V(\overline{\overline{y}}_n) = \frac{n-M}{nN}\overline{S}_b^{\ 2}$ (d) $V(\overline{\overline{y}}_n) = \frac{n-\overline{M}}{nN}\overline{S}_b^{\ 2}$

Q.32 In case of unequal cluster size Mi, formula for $\hat{V}(\overline{\overline{y}}_n)$ is given as

(a) $\hat{V}(\overline{\overline{y}}_n) = \frac{N-n}{nN}s_w^{\ 2}$ (b) $\hat{V}(\overline{\overline{y}}_n) = \frac{N-n}{nN}s_b^{\ 2}$

(c) $\hat{V}(\overline{\overline{y}}_n) = \frac{N-\overline{M}}{nN}s_b^{\ 2}$ (d) $\hat{V}(\overline{\overline{y}}_n) = \frac{N-\overline{M}}{nN}\overline{s}_w^{\ 2}$

Q.33 In case of unequal cluster size Mi, an expression for $\hat{V}(\overline{\overline{y}}_n')$ is written as

(a) $\hat{V}(\overline{\overline{y}}_n') = \frac{N-\overline{M}}{nN}S''^{\ 2}_b$ (b) $\hat{V}(\overline{\overline{y}}_n') = \frac{N-n}{nN}S_b^{\ 2}$

(c) $\hat{V}(\overline{\overline{y}}_n') = \frac{N-n}{nN}\overline{S}_w^{\ 2}$ (d) $\hat{V}(\overline{\overline{y}}_n') = \frac{N-n}{nN}s''^{\ 2}_b$

Q.34 The expression for S''^2_b is wirtten as

(a) $S''^2_b = \frac{1}{N-1}\sum_i^N \left(\frac{Mi}{\bar{M}}\right)^2 (\bar{y}_i - \bar{y}..)^2$

(b) $S''^2_b = \frac{1}{N-1}\sum_i^N \left(\frac{Mi}{N}\right)^2 (\bar{y}_i - \bar{y}..)^2$

(c) $S''^2_b = \frac{1}{n-1}\sum_i^n \left(\frac{Mi}{\bar{M}}\right)^2 (\bar{y}_i - \bar{y}..)^2$

(d) $S''^2_b = \frac{1}{N-1}\sum_{i=1}^N \left(\frac{\bar{M}}{Mi}\right)^2 (\bar{y}_i - \bar{y}..)^2$

Q.35 A consistent estimate of $V(\bar{\bar{y}}_n')$ can be worked out using the expression

(a) $\hat{V}(\bar{\bar{y}}_n') = \frac{N-n}{nN} s''^2_b$

(b) $\hat{V}(\bar{\bar{y}}_n') = \frac{N-n}{nN} s_b^2$

(c) $\hat{V}(\bar{\bar{y}}_n') = \frac{N-n}{nN} \bar{s}_w^2$

(d) $\hat{V}(\bar{\bar{y}}_n') = \frac{N-n}{nN} s'^2_b$

Q.36 Sampling variance of an unbiased estimator $\bar{y}_n'$ is written as

(a) $V(\bar{y}_n') = \frac{N-n}{nN} S'^2_b$

(b) $V(\bar{y}_n') = \frac{N-n}{nN} S_b^2$

(c) $V(\bar{y}_n') = \frac{N-n}{nN} S_b''^2$

(d) $V(\bar{y}_n') = \frac{N-n}{nN} \bar{S}_w^2$

Q.37 Expression for S'^2_b in case of unequal cluster size is written as

(a) $S'^2_b = \frac{1}{N-1}\sum_{i=1}^N \left(\frac{Mi}{\bar{M}}\bar{y}_i - \bar{y}..\right)^2$

(b) $S'^2_b = \frac{1}{N-1}\sum_i^N \frac{Mi}{\bar{M}}(\bar{y}_i - \bar{y}..)^2$

(c) $S'^{2}_{b} = \frac{1}{N-1}\sum_{i}^{N}\left(\frac{Mi}{\bar{M}}\right)^{2}(\bar{y}_{i} - \bar{y}..)^{2}$

(d) $S'^{2}_{b} = \frac{1}{n-1}\sum_{i}^{N}\left(\frac{Mi}{\bar{M}}\right)^{2}(\bar{y}_{i} - \bar{y}..)^{2}$

Q.38 In case of unequal cluster size Mi, an expression to work out an unbiased estimate of $\hat{V}(\bar{y}_{n}')$ is written as

(a) $\hat{V}(\bar{y}_{n}') = \left(\frac{N-n}{nN}\right)s_{b}^{2}$

(b) $\hat{V}(\bar{y}_{n}') = \left(\frac{N-n}{nN}\right)s'^{2}_{b}$

(c) $\hat{V}(\bar{y}_{n}') = \left(\frac{N-n}{nN}\right)\bar{s}_{w}^{2}$

(d) $\hat{V}(\bar{y}_{n}') = \left(\frac{N-n}{nN}\right)\bar{s}'^{2}_{w}$

Q.39 In case of unequal cluster size M i, s'^{2}_{b} is defined as.

(a) $s'^{2}_{b} = \frac{1}{n-1}\sum_{i}^{n}\left(\frac{Mi}{\bar{M}}\bar{y}_{i} - \bar{y}_{n}'\right)^{2}$

(b) $s'^{2}_{b} = \frac{1}{n-1}\sum_{i}^{n}\frac{Mi}{\bar{M}}(\bar{y}_{i} - \bar{y}_{n}')^{2}$

(c) $s'^{2}_{b} = \frac{1}{n-1}\sum_{i}^{n}\left(\frac{Mi}{\bar{M}}\right)^{2}(\bar{y}_{i} - \bar{y}_{n}')^{2}$

(d) $s'^{2}_{b} = \frac{1}{n-1}\sum_{i=1}^{n}\left(\frac{Mi}{\bar{M}o}\bar{y}_{i} - \bar{y}_{n}\right)^{2}$

4.6 Key Answers

4.6.1 True or False

1.	True	2.	True	3.	True
4.	True	5.	False	6.	True
7.	False	8.	True	9.	True
10.	False	11.	True	12.	True
13.	True	14.	True	15.	True

16.	True	17.	False	18.	True
19.	True	20.	True	21.	True
22.	True	23.	True	24.	True
25.	True	26.	True	27.	True
28.	True	29.	True	30.	True
31.	True	32.	True	33.	True
34.	True	35.	True	36.	True
37.	True	38.	True	39.	True

4.6.2 Fill in The Blanks

1. Elementary unit
2. Cluster
3. Cluster
4. Complete
5. One and only one
6. Bias
7. Less
8. Increase
9. $0 \leq \rho_w < 1.0$
10. Unit, cluster
11. Mean of ith cluster
12. Mean of n cluster means
13. Population
14. Within, Population
15. Between, Population
16. Elements, Population
17. Unbiased
18. $V(\overline{\overline{y}}_n) = \frac{N-n}{nN} S_b^2$
19. $\hat{V}(\overline{\overline{y}}n) = \frac{N-n}{nN} s_b^2$

20. Heterogeneous
21. $(M-1)\rho_w$
22. 1 or one
23. M. S_b^2
24. Decrease
25. Biased
26. Unbiased
27. $\frac{(N-1)}{Mo}$
28. $(Mi-\bar{M})(\bar{y}_i-\bar{\bar{y}}_N)$
29. $\left(\frac{N-n}{nN}\right), \left(\frac{N-1}{Mo}\right)$
30. Zero
31. Unbiased
32. $\left(\frac{N-n}{nN}\right)$
33. $(Mi/\bar{M})$
34. $\left(\frac{N-n}{nN}\right)$
35. $\left(\frac{Mi}{\bar{M}}\right)$
36. $\left(\frac{1}{n-1}\right)\sum_{i}^{n}\left(\frac{Mi}{\bar{M}}\bar{y}_i-\bar{y}_n'\right)^2$

4.6.3 Multiple Choice Questions and Answers

1.	d	2.	a	3.	d
4.	d	5.	a	6.	b
7.	b	8.	b	9.	a
10.	a	11.	d	12.	a
13.	a	14.	a	15.	a

16.	b	17.	a	18.	a
19.	a	20.	c	21.	c
22.	d	23.	a	24.	b
25.	b	26.	b	27.	c
28.	b	29.	c	30.	b
31.	b	32.	b	33.	a
34.	a	35.	a	36.	a
37.	a	38.	b	39.	a

5

Systematic Sampling

5.1 True / False

Q.1 Systematic random sampling or systematic sampling, both are different sampling schemes.

Q.2 The pattern usually followed in selecting a systematic sample is a simple pattern involving a constant spacing of units in the sample.

Q.3 In case of systematic sampling, if N = n k where N and n are the population and sample sizes respectively then 'k' is known as 'sampling interval'.

Q.4 In case of systematic sampling with i, n, and k as random start, sample size and sampling interval, the serial number of ________ units in the sample is represented by

y_{ij} = i + (J-1) K, j - 1, 2, 3, ________ , n.

Q.5 The theory of systematic sampling can be studied from the theory of cluster sampling where only one cluster is selected as one sample out of K samples.

Q.6 The sampling variance of the sample mean $\overline{y}_{sys}$ in systematic sampling can be estimated unbiasedly very easily.

Q.7 Linear systematic sampling scheme is one out of the two sampling procedures of select the systematic samples.

Q.8 If N = n k in case of systematic sampling then there is a concept to select one cluster at random out of k clusters each of size n and it is treated as a systematic sample.

Q.9 In systematic sampling if N = n k, then there is a concept that there is n strata each of size k and only one unit from each stratum is selected to constitute a systematic sample.

Q.10 Systematic sampling can be regarded as the particular case of cluster sampling.

Q.11 Systematic sampling can be regarded as the particular case of stratified random sampling where only one unit is selected from each of n strata.

Q.12 When $N \neq n\,k$, then linear systematic sampling scheme can provide all the systematic samples each of the same size n.

Q.13 When $N \neq n\,k$, then circular systematic sampling scheme provides a procedure to prepare all possible systematic samples each of same size n.

Q.14 In case $N \neq n\,k$, to adopt the circular systematic sampling, the appropriate sampling interval k will be an interval nearest to $K = \frac{N}{n}$.

Q.15 The systematic sample mean $\overline{y}_{sys}$ defined as $\overline{y}_{sys} = \frac{1}{n}\sum_{i=1}^{n} y_{ij}$, gives an unbiased estimate of population mean $\overline{Y}_N$.

Q.16 An expression to work out the sampling variance of the sample mean $\overline{y}_{sys}$ is written as $V(\overline{y}_{sys}) = \frac{(K-1)}{K} Sc^2$

Where $Sc^2 = \frac{1}{(K-1)} \sum_{i=1}^{k} (\overline{y}_i - \overline{y}_N)^2$ denotes the mean square of the systematic sample means $\overline{y}_i$ ($i = 1, 2, \ldots\ldots, k$)

Q.17 The sampling variance $V(\overline{y}_{sys})$ can be reduced by increasing the sample size n because $V(\overline{y}_{sys})$ is not the inverse function of n.

Q.18 $V(\overline{y}_{sys})$ can also be expressed as $V(\overline{y}_i) = \frac{N-1}{n\,k} s^2 - \frac{(n-1)}{n} \overline{S}_w^{\,2}$

where $\overline{S}_w^{\,2} = \frac{1}{k}\sum_{i}^{k} S_i^2 = \frac{1}{k}\sum_{i}^{k}\left[\sum_{j=1}^{n} (\overline{y}_{ij} - \overline{y}_i)^2\right]$

Q.19 The two terms Sc^2 and $\overline{S}_w^{\,2}$ are respectively known as "between and within" samples mean squares.

Q.20 To reduce the sampling variance $V(\overline{y}_{sys})$, the samples must not be as heterogeneous as possible within themselves.

Q.21 If population units are of the random nature then systematic sampling scheme and S.R.S.W.O.R. scheme, both are equally efficient.

Q.22 If the sampling units in the population follow a linear trend then the relation among sampling variances of the sample means under (i) systematic sampling (ii) stratified random sampling with one unit per stratum and (iii) S R S W O R exists as-

$Vstr \geq Vsys \geq Vrandom$

Q.23 If the population units have a linear trend then systematic sampling is found more efficient than the usual stratified random sampling.

Q.24 When the sampling interval k is equal to the period of the curve or an integral multiple of the period, the value of every unit selected in the systematic sample will be equal.

Q.25 In situation expressed in Q.24, the sampling variance of the sample mean is equal to the population variance σ^2.

Q.26 If the sampling interval k is equal to an odd multiple of half period, the every systematic sample mean is equal to the population mean.

Q.27 In situation described in Q.26, the sampling variance of the sample mean is equal to zero, and systematic sampling is best of all sampling procedures.

Q.28 When periodic effect is suspected on sampling units or is not known with surity then S.R.S.W.O.R. or stratified random sampling will not be preferable.

Q.29 In case of natural populations, it has been found empirically that systematic sampling performs better than other sampling schemes.

5.2 Fill in the Blanks

Q.1 Systematic sampling scheme is also known as ____________ sampling scheme.

Q.2 In systematic sampling scheme, if population size N is expressed as N = n k then 'k' is known as ____________.

Q.3 In systematic sampling scheme if the units of the sample are represented as $i, i + k, i + 2k.$ ____________ $i + (n-1) k$ then 'i' is known as ____________

Q.4 Systematic sampling scheme can be regarded as a particular case of ____________ where only one cluster is selected at random out of k clusters to constitute a sample.

Q.5 In systematic sampling scheme an ____________ estimate of sampling variance of the sample mean $\overline{y}_{sys}$ can not be obtained.

Q.6 In systematic sampling where N = n k, and we use the concept that there are n strata each of size k, then number of units selected from each stratum is equal to ___________.

Q.7 If $N \neq k\,n$ then to list-out, all the possible systematic samples each of size n, the name of sampling technique adopted is ___________.

Q.8 In case $N \neq k\,n$, to adopt circular systematic sampling, the value of sampling interval k will be an integer nearest to ___________.

Q.9 An expression to work out the sampling variance of the sample mean $\overline{y}_{sys}$ is written in term of Sc^2 as $V(\overline{y}_{sys}) =$ ___________.

Q.10 The $V(\overline{y}_{sys})$ cannot be reduced by increasing the sample size n because $V(\overline{y}_{sys})$ is not the ___________ function of n.

Q.11 The term S^2_C used in the expression to workout the value of $V(\overline{y}_{sys})$ is known as mean square ___________ cluster means.

Q.12 The term $\overline{S}^2_w = \frac{1}{k}\sum_{i}^{k} S_i^2$ is known as the mean square ___________ the sample.

Q.13 To reduce the sampling variance $V(\overline{y}_{sys})$ the systematic sample must be as as ___________ possible within themselves.

Q.14 If population units are of the random nature then systematic sampling and S.R.S.W.O.R scheme are ___________ efficient.

Q.15 In case of population following linear trend, the relation among sampling variance of sample mean $\overline{y}_{sys}$ under (i) Systematic sampling (ii) Stratified random sampling with one unit per stratum and (iii) S.R.S.W.O.R. exists as follows. ___________

Q.16 In case of linear trend being followed in population, the systematic sampling is found ___________ efficient than usual stratified random sampling.

Q.17 When sampling interval k is equal to the period of curve or an integral multiple of the period, the value of every unit selected in the systematic sample will be ___________

Q.18 In situation expressed in Q.17, the $V\left(\overline{y}_{sys}\right)$ will be equal to the ___________

Q.19 If sampling interval k is equal to an odd multiple of half period, the every systematic sample mean will be equal to the ___________

Q.20 In situation described in Q.19, the $V\left(\overline{y}_{sys}\right)$ will be equal to ________

Q.21 In case of situtation described in Q.19, out of all the sampling schemes, the best sampling scheme will be the ________

Q.22 When the periodic effect on sampling uint of the population is suspected or is not known with surity then the preferable sampling schemes will be either ________ or ________

Q.23 In case of natural population, the sampling scheme better than other sampling schemes is the ________

5.3 Multiple Choice Questions and Answers

Q.1 Systematic sampling is also known as-

(a) Cluster sampling

(b) Systematic random sampling

(c) Systematic cluster sampling,

(d) Systematic sub-sampling

Q.2 In case of systematic sampling if N is expressable as N = n k then K is known as.

(a) Sampling constant (b) Sampling interval

(c) Sampling difference (d) Sampling fraction

Q.3 If (*i*, n, k) denote random start, sample size and sampling interval respectively then the serial number of units included in the systematic sample is represented as

(a) i + (j -1) k, j = 1, 2, 3, ... , n

(b) i + (j +1) k, j = 1, 2, 3, ... , n

(c) i + (J -1) k, j = 1, 2, 3, ... (n + 1)

(d) i + (J -1) k, j = 1, 2, 3, ... , (n - 1)

Q.4 If K samples are treated as cluster when N population units are arranged in n rows and k columns then theory of systematic sampling can be studied from the theory of

(a) Cluster sampling

(b) Two way systematic sampling

(c) Stratified random sampling

(d) Two stage sampling

Q.5 Systematic sample mean estimator $\overline{y}_{sys}$ provides -
(a) A consistent estimate of population mean.
(b) A biased estimate of population mean.
(c) An unbiased estimate of population mean.
(d) A sufficient estimate of population mean.

Q.6 In systematic sampling if $N = n\,k$ and we use the concept that there are n strata each of size k then number of units drawn from each stratum to constite a systematic sample is-

(a) Equal to k (b) Equal to one

(c) Equal to $\frac{n}{k}$ (d) Equal to $\frac{N}{k}$

Q.7 If $N \neq n\,k$, then to list out all possible systematic samples, each of size n, the sampling scheme adopted is known as-

(a) Linear systematic sampling
(b) Circular systematic sampling
(c) Cluster sampling
(d) None of the above

Q.8 In case $N \neq n\,k$, to adopt circular systematic sampling the value of sampling interval k will be an integer nearest to

(a) $\frac{n}{k}$ (b) $\frac{N}{n}$

(b) $\frac{n}{N}$ (d) $\frac{k}{n}$

Q.9 The sampling variance of the sample mean $\overline{y}_{sys}$ can be worked out from the expression-

(a) $V(\overline{y}_{sys}) = \frac{k}{k-1} Sc^2$ (b) $V(\overline{y}_{sys}) = \frac{k-1}{k} Sc^2$

(c) $V(\overline{y}_{sys}) = \frac{nk-1}{k} Sc^2$ (d) $V(\overline{y}_{sys}) = \frac{nk}{n-1} Sc^2$

where Sc^2 is mean square between sample means.

Q.10 The term Sc^2 used in the expression of $V(\overline{y}_{sys})$ is known as

(a) Mean square between the sample means
(b) Mean square within the sample
(c) Population means square
(d) Mean square of the individual units in the population

Q.11 The term $\overline{S}w^2$ used in the expression of $V(\overline{y}_{sys})$ and defined as $\overline{S}w^2 = \frac{1}{k}\sum_{i=1}^{K} S_i^2$ is known as -

(a) Mean square between sample means
(b) Mean square within sample
(c) Error mean square
(d) Average population mean square

Q.12 To reduce the sampling variance $V(\overline{y}_{sys})$, the systematic samples must be-

(a) As heterogeneous as possible within themselve.
(b) As homogeneous as possible between themselve.
(c) As heterogeneous as possible between themselves.
(d) As homogeneous as possible within themselves.

Q.13 If population units are of the random nature then-

(a) Systematic sampling and S.R.S.W.O.R. are equally efficient.
(b) Systematic sampling is less efficient them S.R.S.W.O.R.
(c) Systematic sampling is more efficient than S.R.S.W.O.R.
(d) Nothing can be said with surity.

Q.14 In case of population following the linear trend, the relation among sampling variance of the sample mean under: (i) Systematic sampling (ii) Stratified random sampling and (iii) S. R. S. W. O. R exists as follows

(a) Vstr $\geq$ Vsys $\geq$ V random (b) Vstr $\leq$ Vsys $\leq$ V random
(c) Vstr $\leq$ Vsys and Vsys $= 0$ (d) Vstr $\geq$ Vrom and Vsys $= 0$

Q.15 If population units follow linear trend the efficiency (e) of systematic sampling w.r. to stratified random sampling will be as-

(a) $e \geq 1.0$ (b) $e \leq 1.0$
(c) $e = 1$ (d) $e = 0.5$

Q.16 When the sampling interval k is equal to the prediod of the curve or intergral multiple of period in case of periodic trend is followed by the population units then

(a) The value of every unit included in the sample will be same.
(b) The estimate of population mean will be equal to the value of any unit included in the sample.
(c) The sample mean will be equal to the value of any unit included in the sample.
(d) All the above three results expected are true.

Q.17 In case of situation expressed in Q.16, the sampling variance of the sample mean will be equal to-

(a) Population variance $\sigma 2$ (b) $\frac{\sigma 2}{N}$

(c) $\frac{\sigma 2}{n}$ (d) $\frac{\sigma^2}{n^2}$

Q.18 If the sampling interval k is equal to an odd multiple of half period then every sample mean $(\overline{y}_{sys})$ will be

(a) Equal to population mean.
(b) Grater than population mean.
(c) Less than population mean.
(d) Nothing can be said with surity.

Q.19 In situation described in Q.16 the sampling variance $V(\overline{y}_{sys})$ will be as

(a) $V(\overline{y}_{sys}) = 0$ (b) $V(\overline{y}_{sys}) > 0$

(c) $V(\overline{y}_{sys}) \leq \sigma 2$ (d) $V(\overline{y}_{sys}) = \frac{\sigma 2}{n}$

Q.20 In situation described in Q.18 we have the -

(a) Systematic sampling and stratified random sampling will be equaly efficient
(b) Systematic sampling will be best among all sampling schemes.
(c) Sampling variance of systematic sample mean will be equal to zero.
(d) Both the results (b) and (c) are true.

Q.21 When periodic effect is suspected on sampling units or is not known with surity then-

(a) S.R.S.W.O.R. is preferable.
(b) Strafified random sampling is preferable
(c) Systematic sampling is preferable.
(d) Nothing can be said about the above.

Q.22 In case of natural population, it has been found empirically that

(a) S.R.S.W.O.R. performs better than systematic sampling
(b) Stratified reandom sampling performs better
(c) Systematic sampling performs better than other sampling schemes.
(d) None of the above

5.4 Key Answers

5.4.1 True / False

1.	False	2.	True	3.	True
4.	True	5.	True	6.	False
7.	True	8.	True	9.	True
10.	True	11.	True	12.	False
13.	True	14.	True	15.	True
16.	True	17.	False	18.	True
19.	True	20.	False	21.	True
22.	False	23.	False	24.	True
25.	True	26.	True	27.	True
28.	False	29.	True		

5.4.2 Fill in the Blanks

1. Random Systematic
2. Sampling Interval
3. Random start
4. Cluster sampling
5. Unbiased
6. One

7. Circular Systematic sampling
8. $\frac{N}{n}$
9. $V(\bar{y}_{sys}) = \frac{k-1}{k} Sc^2$
10. Inverse
11. Between
12. Within
13. Heterogeneous
14. Equally
15. Vst $\leq$ Vsys $\leq$ Vrand
16. Less
17. Equal or same
18. Population variance σ^2
19. Population mean
20. Zero
21. Systematic sampling
22. S.R.S.W.O.R. or stratified Random Sampling
23. Systematic sampling

5.4.3 Multiple Choice Questions and Answers

1.	b	2.	b	3.	a
4.	a	5.	c	6.	b
7.	b	8.	b	9.	b
10.	a	11.	b	12.	a
13.	a	14.	b	15.	b
16.	d	17.	a	18.	a
19.	a	20.	d	21.	c
22.	c				

6

Ratio Product and Regression Methods of Estimation

6.1 True / False

Q.1 Informations available on an auxiliary variable X which is highly correlated with the study variable y, can be utilized in improving the sampling design for the study variable y.

Q.2 Information on auxiliary variable X can be used for stratification of population for the study character y.

Q.3 In probability proportional to size (P P S) sampling, an auxiliary information X (size) can be utilized to improve the sampling design to be used for study variable Y.

Q.4 An aggregate value (X) of all sampling units of an auxiliary character (x) available can not be utilized at estimation stage to get an improved estimate of population parameter (Mean, Total) of the study character y.

Q.5 The ratio, product and regression methods of estimation never utilize the informations available on an auxiliary variate x, to get improved estimate of parameters of study variable y.

Q.6 One of the main aims of using informations available on an auxiliary variable x is to get an increased precision of an estimate of population mean $\overline{y}_N$.

Q.7 An auxiliary variate x is also known as co-variate or predictor variate.

Q.8 The study variate y is also known as response variate or regressed variate.

Q.9 The co-rrelationship existing between auxiliary variate x and dependent variate y can not be utilized t o improve the sampling design used for study variate y.

Q.10 Informations on auxiliary variable x can be taken from prẹvious census or records.

Q.11 During performing the current survey for study variate y, the information collected on a larger sample (say n' > n) on auxiliary variate x in a easy and low cost can be used to estimate $\overline{X}_N$.

Q.12 When the correlation coefficient ρ_{yx} is positive and high and line passes through origin then ratio method of estimation can not be utilized efficiently.

Q.13 When the correlation coefficient ρ_{yx} is positive and high, then product method of estimation can be utilized successfully.

Q.14 To estimate the population mean $\overline{Y}_N$, the ratio estimator $\overline{y}_R$ is defined as $\overline{y}_R = \frac{\overline{x}n}{\overline{y}n}\overline{X}_N$

Q.15 When $\overline{X}_N$ is not known then the ratio estimator for $\overline{Y}_N$ is defined as $\overline{y}_{Rd} = \frac{\overline{y}n}{\overline{x}n}\overline{x}_{n'}$.

Q.16 The population ratio $R_N = (\overline{Y}_N / \overline{X}_N)$ is estimated by the sample estimator Rn defined as Rn = $\frac{\overline{x}n}{yn}$.

Q.17 The ratio of area under wheat cultivation (Y) to the area under paddy cultivation (X) in a state can be defined as $Ra = y_a / x_a$ where (y_a / x_a) denote the area cultivtion of the wheat and paddy respectively in some selected districts of the state.

Q.18 The bias of ratio estimator $\overline{y}_R$ is defined as $B(\overline{y}_R) = (\overline{Y}_N - E(\overline{y}_R))$

Q.19 The expression for bias of ratio estimator $(\overline{y}_R)$ is obtained as $B(\overline{y}_R) = \text{cov}(R_{n,}\ \overline{x}n)$

Q.20 The expression to work out the first order approximate value of bias $(\overline{y}_R)$ is given as

$$B_1(\overline{y}_R) = \frac{(N-n)}{n\,N}\overline{Y}_N(Cx^2 - \rho\,Cx\,Cy)$$ in usual notations.

Q.21 If $Cx = Cy = C$, then $B_1(\overline{y}_R)$ can be exprssed as

$$B_1(\overline{y}_R) = \left(\frac{N-n}{n\,N}\right)\overline{Y}_N C^2(1-\rho)$$

Q.22 When ρ will be negative and high, $B_1(\overline{y}_R)$ will be high.

Q.23 When ρ will be positive and high, then bias $(\overline{y}_R)$ will be large.

Q.24 In case of perfect positive correlation ρ_{yx}, $cx = cy$, the $B_1(\overline{y}_R)$ will be equal to zero.

Q.25 If regression line of y on x is linear and passes through origin, then $B_1(\overline{y}_R)$ will be equal to zero.

Q.26 The upper limit of bias $(\overline{y}_R)$ is given as upper $B(\overline{y}_R) = \sigma_{Rn}\sigma_{\overline{x}n}$.

Q.27 The upper limit of bias to the estimators Rn and $\hat{Y}R$ is written as

upper $B(\overline{y}_{Rn}$, or $\hat{y}_R) = \sigma_{Rn}\ \sigma\overline{x}_n$

Q.28 The M. S. E. $(\overline{y}_R)$ is defined as $M(\overline{y}_R) = E(\overline{y}_R - Y_N)^2$

Q.29 The M. S. E. $(\overline{y}_R)$ can also be defined as

$$M(\overline{y}_R) = \left(\frac{N-n}{nN}\right)\left(S_y{}^2 + R_N{}^2 S_x{}^2 + 2R_N\,Sxy\right)$$

Q.30 The first order approximation to M. S. E. $(\overline{y}_R)$ is expressed as

$$M_1(\overline{y}_R) = \left(\frac{N-n}{n\,N}\right)\overline{y}_N{}^2\,(C_y^2 + C^2x + 2\rho\,CxCy)$$

Q.31 The first order approximation to M. S. E. $(\overline{y}_R)$ can also be expressed in terms of C. V. S of x and y as-

$$B_1(\overline{y}_{Rd}) = \left(\frac{n'-n}{n\,n'}\right)\overline{Y}_N(Cx^2 + \rho Cx\,Cy)$$

Q.32 The first order approximation to the bias of the estimator $\overline{y}_{Rd}$ can be written as

$$M_1(\overline{y}_{Rd}) = \left(\frac{n' - n}{n\,n1}\right)\overline{Y}_N^2\,(C_y^2 + Cx^2 + 2\rho Cx\,Cy) + \frac{N - n1}{n\,N}(\overline{Y}_N^2\,C_y^2)$$

Q.33 Under certain conditions, the ratio estimator $\overline{y}_R$ is found more efficient then the sample mean $\overline{y}_n$.

Q.34 Under large sample for ratio estimator $\overline{y}_R$ being more efficient than the sample mean $\overline{y}_n$, the condition is $\rho_{yx} < \frac{1}{2}\frac{Cx}{Cy}$

Q.35 Under large sample and $Cx = Cy$, the ratio estimator is found more efficient than the sample mean if $\rho > \frac{1}{2} = 0.5$

Q.36 An estimate of $B_1(\overline{y}_R)$ can be obtained from an expression written as

$\hat{B}_1(\overline{y}_R) = \frac{N - n}{n\,N}\overline{y}_n\,(Cx^2 - r\,C_x C_y)$ where r, C_x, C_y are sample value.

Q.37 $M_1(\overline{y}_R)$ can be estimated through an expression written as

$$\hat{M}_1(\overline{y}_R) = \left(\frac{N - n}{n\,N}\right)(sy^2 + Rn^2 s^2 x + 2Rn\,sxy)$$

Q.38 $B_1(\overline{y}_{Rd})$ can be estimated through an expression written as

$$\hat{B}_1(\overline{y}_{Rd}) = \left(\frac{n' - n}{n\,n1}\right)\overline{y}_n(Cx^2 - r\,Cx\,Cy)$$

Q.39 An expression to workout the estimate of $M_1(\overline{y}_{Rd})$ can be written as

$$\hat{M}_1(\overline{y}_{Rd}) = \left(\frac{n' - n}{n\,n1}\right)\overline{y}_n{}^2(cy^2 + cx^2 + 2r\,cx\,cy) + \left(\frac{N - n}{n\,N}\right)\overline{y}_n{}^2 c_y{}^2$$

Q.40 An unbiased ratio type estimator $\overline{y}_{RU}$ suggested by Hartley and Ross (1954) is defined as

$$\overline{y}_{RU} = \overline{r}_n\;\overline{x}_n + \frac{n(N - 1)}{(n - 1)N}(\overline{y}_n - \overline{r}_n\;\overline{x}_n)$$

where $\overline{r}_n = \frac{1}{n}\sum_i^n r_i$, $r_i = \frac{y_i}{x_i}, (i = 1, 2, \ldots\ldots\ldots, n)$

Q.41 The sampling variance of the estimator $\overline{y}_{RU}$ can be worked out from an expression given as

$$V_1(\bar{y}_{RU}) = \frac{1}{n}(S_y^{\,2} + \bar{r}_N^2 S_x^2 - 2\bar{r}_N\, Sxy) \text{ where } \bar{r}_N = \frac{1}{N}\sum_i^N r_i$$

Q.42 If ρ_{yx} is negative and high then an alternative to ratio etimator known as product estimator is recommended.

Q.43 The product estimator is defined as $\bar{y}_\rho = \dfrac{\bar{y}_n \bar{x}_n}{\bar{X}_N}$

Q.44 If the value of population mean $\bar{X}_N$ is not known in prior then an other product estimator $\bar{y}_{Pd}$ is recommended and is defined as $\bar{y}_{Pd} = \dfrac{\bar{y}_n \bar{x}_n}{\bar{x}_{n'}}$ where $\bar{x}_{n'}$ is a sample mean of size n' >n.

Q.45 The product estimator $\bar{y}_\rho$ is a biased estimator of population mean $\bar{y}_N$ and its amount of bias is expressed as

$$B(\bar{y}_\rho) = \left(\frac{N-n}{nN}\right)\frac{Sxy}{\bar{X}n}$$

Q.46 For a large sample size, the relation between relative bias and can be expressed as $\dfrac{B(\bar{y}_\rho)}{y_N} < Cx\,Cy$

Q.47 In case of large samples, an expression for $(\bar{y}\rho)$ to the first order approximation is written as

$$V(\bar{y}_\rho)\left(\frac{N-n}{nN}\right)(S_y^{\,2} + R_N^{\,2}\, Sx^2 + 2R_N\, Sxy)$$

Q.48 The bias $B_1(\bar{y}_\rho)$ can be estimated through an expression given as

$$\hat{B}_1(\bar{y}_\rho) = \frac{N-n}{nN}\frac{sxy}{\bar{x}_n}$$

Q.49 The sampling variance $V_1(\overline{Y}\rho)$ can be estimated through an expression written as

$$\hat{V}_1(\overline{Y}\rho) = \frac{N-n}{nN}(Sy^2 + R_n^2\, Sx^2 + 2R_N\, Sxy)$$

Q.50 For the product estimator $\bar{y}_\rho$ to be more efficient than the sample mean estimator $\bar{y}n$, the following condition must be satisfied. $\rho < \dfrac{1}{2}\dfrac{Cx}{Cy}$

Q.51 If $Cx = Cy$ and $\rho < \frac{1}{2}$ (or 0.5) then product estimator $\overline{y}_{\rho}$ will always be more efficient than sample mean $\overline{y}_{n}$.

Q.52 If the linear regression line of Y on X does not pass though origin then the regression estimator is quite effective.

Q.53 If the line of Y on X is linear, ρ_{yx} is positive and high but line does not pass through origin then ratio estimator is quite effective.

Q.54 The difference estimator to estimate the population mean $\overline{Y}_{N}$, is defined as $\overline{y}_{D} = \overline{y}_{n} + K(\overline{X}_{N} - \overline{x}_{n})$ where K is some constant.

Q.55 The difference estimator $\overline{y}_{D}$ gives a biased estimate of population mean $\overline{y}_{N}$.

Q.56 The difference estimator is not analogus to regression estimator.

Q.57 If the value of K in difference estimator $\overline{y}_{D}$ is replaced by the sample regression coefficient by x then, it reduces to the regression estimator defined as $\overline{y}_{lr} = \overline{y}_{n} + b_{yx}(\overline{X}_{N} - x_{n})$

Q.58 The regression estimator $\overline{y}_{lr}$ provides a biased estimate of population mean $\overline{y}_{N}$.

Q.59 An approximate compact expression for bias $(\overline{y}_{lr})$ can be written as $B_{1}(\overline{y}_{lr}) = -\text{Covariance}(b_{yx}, \overline{x}_{n})$

Q.60 An approximate variance expression of the estimator $\overline{y}_{lr}$ is written as

$$V_{1}(\overline{y}_{lr}) = \left(\frac{N-n}{nN}\right) Sy^{2}(1+\rho_{yx}^{2})$$

Q.61 In case of correlation coefficient ρ_{yx} being is high (either positive or negative), the $V_{1}(\overline{y}_{lr})$ will be high.

Q.62 For ρ_{yx} being perfect positive or perfect negative, the variance $V_{1}(\overline{y}_{lr})$ becomes equal to zero.

Q.63 An expression to estimate the $V_1(\overline{y}_{lr})$ can be written as

$$\hat{V}_1(\overline{y}_{lr}) = \left(\frac{N-n}{nN}\right) sy^2 (1 - r_{xy}{}^2)$$

Q.64 In case of ρ_{yx} being not equal to zero, the regression estimator $\overline{y}_{lr}$ will always be more precise than the sample estimator $\overline{y}_n$ for population mean $\overline{y}_N$.

Q.65 If the value of ρ_{yx} is exactly equal to zero then both the estimators $\overline{y}_{lr}$ and $\overline{y}_n$ will be equally efficient.

Q.66 If regression line of y on x is linear but does not pass through origin then ratio estimator will always be inferior to regression estimator to estimate $\overline{Y}_N$.

Q.67 If $R_N = \beta_{yx}$ then both the estimators $\overline{y}_R$ and $\overline{y}_{lr}$ will be equally efficient.

Q.68 If the line of regression of y on x is linear and it passes through origin then $R_N \neq \beta_{yx}$

Q.69 If the relation between (y, x) is linear, the line passes through origin and $V(y_i/x_i) \, \alpha \, x_i$ about this line then ratio estimator $\overline{y}_R$ is best linear unbiased estimator (B. L. U. E.) for population mean $\overline{y}_N$.

Q.70 Let y_{hi} (h=1, 2 .... k, j =1, 2, ... N_h) denotes the Jth unit of hth stratum such that $\sum_{h=1}^{n} N_h = N$ then the separate ratio estimator $\overline{y}_{RS}$ is defined as $\overline{y}_{RS} = \sum_{h=1}^{n} Wh\,\overline{y}_{Rh}$ for $\overline{y}_{Rh} = \frac{\overline{y}_{nh}}{\overline{x}_{nh}} \overline{X}_{Nh}$

Q.71 The expression for $V(\overline{y}_{RS})$ can be written as

$$V(\overline{y}_{RS}) = \sum_{h=1}^{k} W_h^2 \left(\frac{N_h - n_h}{n_h N_h}\right) \left\{S^2{}_{hy} + R^2{}_{Nh} S^2 hx + 2R_{Nh} \; S_{hxy}\right\}$$

Q.72 The expression written in Q.(70) gives good results only for $n_h > 30$ for all h = 1, 2, ...k,

Q.73 If n_h is small and k is large then bias $(\overline{y}_{RS})$ will be negligible w. r. to . the S. E. $(\overline{y}_{RS})$.

Q.74 The combined ratio estimator $\overline{y}_{RC}$ in stratified random sampling is defined as.

$$\overline{y}_{RC} = \frac{\overline{y}_w}{x_w}\overline{X}_N \text{ for } \overline{y}_W = \sum_{h=1}^{k} W_h\ \overline{y}_{nh} \text{ and } \overline{x}_w = \sum_{h=1}^{k} W_h \overline{x}_{nh}$$

Q.75 The estimator $\overline{y}_{RS}$ has much less risk subject to its bias than that of the estimator $\overline{y}_{RC}$.

Q.76 If the stratum ratio R_h in $V(\overline{y}_{RS})$ becomes equal to the population ratio R then $V(\overline{y}_{RS})$ and $V(\overline{y}_{RC})$ take the same general form.

Q.77 When R_h varies from stratum to stratum and $n_h > 30$, h = 1, 2, ..., k, then $\overline{y}_{RS}$ is found more precise than $\overline{y}_{RC}$

Q.78 If $n_h < 30$, h = 1, 2, ..., k then $\overline{y}_{RC}$ is recommended in relation to $\overline{y}_{RS}$ in practice.

Q.79 The separate regression estimator $\overline{y}_{lrs}$ for the population mean $\overline{Y}_N$ is defined as

$$\overline{y}_{lrs} = \sum_{h=1}^{k} W_h\ \overline{y}_{lrh} \text{ where } \overline{y}_{lrh} = \overline{y}_{nh} + b_h\ (\overline{X}_{Nh} - \overline{x}_{nh})$$

is the regression estimator for h[th] stratum mean $\overline{Y}_{Nh}$

Q.80 The estimator $\overline{y}_{lrs}$ holds good when true regression coefficient β_h varies from stratum to stratum.

Q.81 The combined regression estimator $\overline{y}_{lrc}$ is defined as $\overline{y}_{lrc} = \overline{y}_w + b(\overline{X}_N - \overline{x}_w)$ where $(\overline{y}_w, \overline{x}_w)$ are weighted sample means in stratified random sampling and b is the sample regression coefficient which is same for all the strata.

Q.82 The estimator $\overline{y}_{lrc}$ recommended only when β_h is assumed to be same for all the strata.

Q.83 When the values of b_h and b are assumed to be known in prior then both the estimators $\overline{y}_{lrs}$ and $\overline{y}_{lrc}$ will be unbiased for the parameter $\overline{y}_N$.

Q.84 When sample sizes $n_h < 30$ for all strata then the estimator $\overline{y}_{lrs}$ will be liable to have more bias than that of $\overline{y}_{lrs}$.

Q.85 When there is some curvilinearity in regression than the estimator $\overline{y}_{lrs}$ is safer to be applied unless $n_h > 30$ for all strata.

6.2 Fill in the Blanks

Q.1 For improving the sampling design : of study variable y, the variable highly correlated with y is the variable x which is known as _________

Q.2 For stratification of population for the study character y, the other variable on which information available can be utilized is known as _________

Q.3 An aggregate value of all the sampling units available of an _________ variable can be utilized at estimation stage to get an improved estimate of the study variable y.

Q.4 The ratio, product and regression estimators utilize the information available on an _________ variable to get an improved estimate of study variable y.

Q.5 One of the main aims of using an auxiliary variable x reated to study variable y is to get an increased _________ of estimate of $\overline{Y}_N$.

Q.6 Auxiliary variate x is also known as _________ or _________ variate.

Q.7 The study variate y is also known as _________ variate or _________ variate.

Q.8 Informations on auxiliary variate x can be taken from previous _________ or record.

Q.9 When the correlation coefficient ρ_{yx} is positive and high then the method of estimation used for $\overline{Y}_N$ is known as _________

Q.10 Let $(\overline{x}_n, \overline{y}_n, \overline{X}_N, \overline{Y}_N)$ be the sample means and population means of the auxiliary variate x and study variate y then the ratio estimator $\overline{y}_R$ for $\overline{Y}_N$ is defined as $\overline{y}_R$ = ...

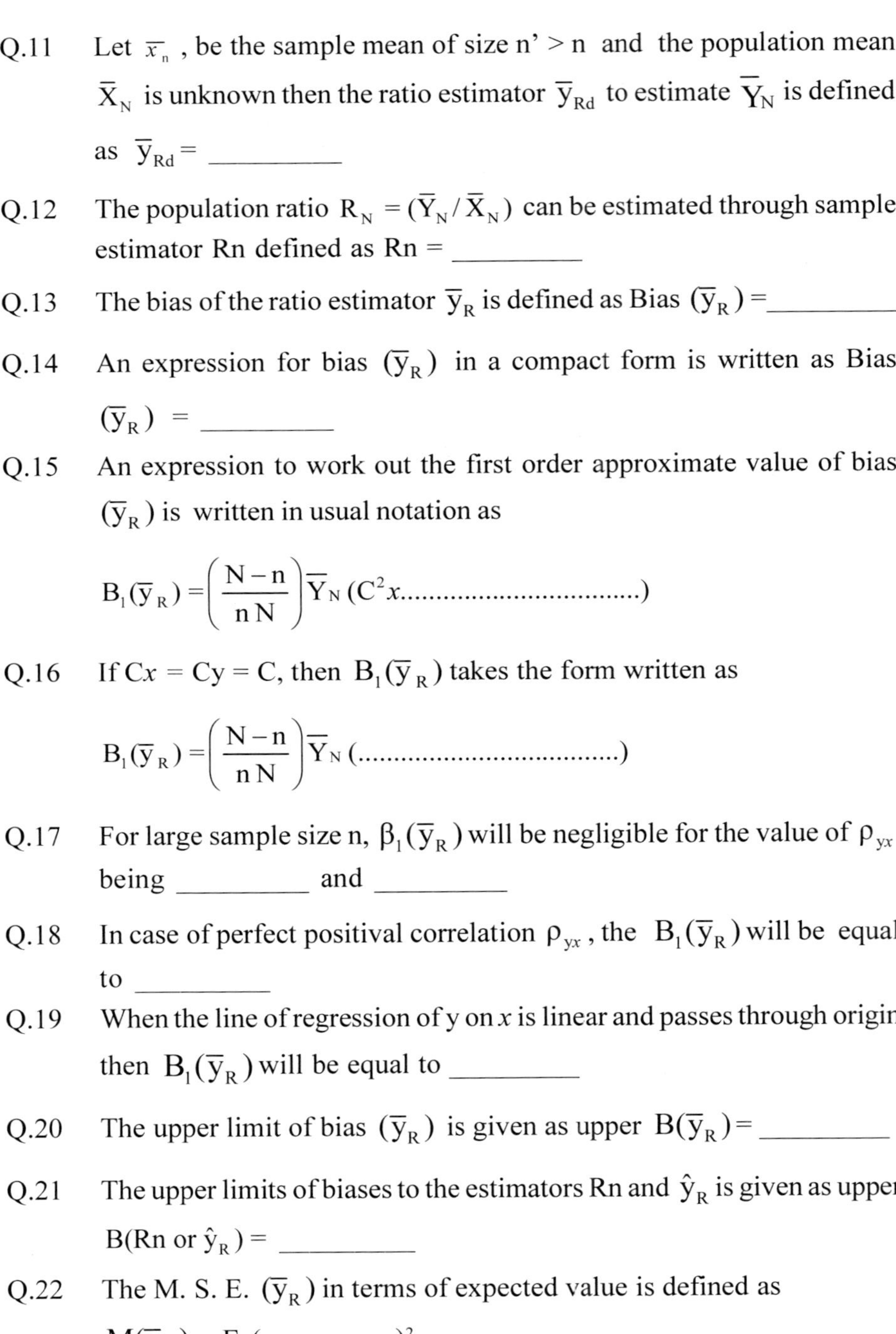

Q.11 Let $\bar{x}_{n}$, be the sample mean of size n' > n and the population mean $\bar{X}_N$ is unknown then the ratio estimator $\bar{y}_{Rd}$ to estimate $\bar{Y}_N$ is defined as $\bar{y}_{Rd} =$ ________

Q.12 The population ratio $R_N = (\bar{Y}_N / \bar{X}_N)$ can be estimated through sample estimator Rn defined as Rn = ________

Q.13 The bias of the ratio estimator $\bar{y}_R$ is defined as Bias $(\bar{y}_R) =$________

Q.14 An expression for bias $(\bar{y}_R)$ in a compact form is written as Bias $(\bar{y}_R)$ = ________

Q.15 An expression to work out the first order approximate value of bias $(\bar{y}_R)$ is written in usual notation as

$$B_1(\bar{y}_R) = \left(\frac{N-n}{nN}\right)\bar{Y}_N\,(C^2x.................................)$$

Q.16 If $Cx = Cy = C$, then $B_1(\bar{y}_R)$ takes the form written as

$$B_1(\bar{y}_R) = \left(\frac{N-n}{nN}\right)\bar{Y}_N\,(.....................................)$$

Q.17 For large sample size n, $\beta_1(\bar{y}_R)$ will be negligible for the value of ρ_{yx} being ________ and ________

Q.18 In case of perfect positival correlation ρ_{yx} , the $B_1(\bar{y}_R)$ will be equal to ________

Q.19 When the line of regression of y on x is linear and passes through origin then $B_1(\bar{y}_R)$ will be equal to ________

Q.20 The upper limit of bias $(\bar{y}_R)$ is given as upper $B(\bar{y}_R) =$ ________

Q.21 The upper limits of biases to the estimators Rn and $\hat{y}_R$ is given as upper B(Rn or $\hat{y}_R$) = ________

Q.22 The M. S. E. $(\bar{y}_R)$ in terms of expected value is defined as $M(\bar{y}_R) = E$ (________$)^2$

Q.23 The sampling variance $V(\overline{y}_R)$ can also be defined as

$$V(\overline{y}_R) = E\ (\_\_\_\_\_\_\_\_\_)^2$$

Q.24 The first order approximation to M. S. E. $(\overline{y}_R)$ is expressed as

$$M_1(\overline{y}_R) = \left(\frac{N-n}{nN}\right)\ (\_\_\_\_\_\_\_\_\_).$$

Q.25 The first order approximation to M. S. E. $(\overline{y}_R)$ can also be expressed in terms of C. V. 's of x and y as.

$$M_1(\overline{y}_R) = \frac{N-n}{nN}\overline{Y}_N^{\,2}\ (\_\_\_\_\_\_\_\_\_).$$

Q.26 The first order approximation to the bias $(\overline{y}_{Rd})$ can be written as

$$B_1(\overline{y}_{Rd}) = \left(\frac{n'-n}{n\,n_1}\right)\overline{Y}_N\,(Cx^2 + \ldots\ldots\ldots)$$

Q.27 For the ratio estimator $\overline{y}_R$ being more efficient than the sample mean $\overline{y}_n$ in large sample approximation, the condition is $\rho_{yx} >$ ....

Q.28 Under large sample approximation and Cx = Cy the ratio estimator $\overline{y}_R$ is found more efficient than the sample mean under the condition $\rho. >$............

Q.29 $B_1(\overline{y}_{Rd})$ can be estimated through an expression written as

$$\hat{B}_1(\overline{y}_{Rd}) = \left(\frac{n'-n}{n\,n'}\right)\overline{y}_n\ (\_\_\_\_\_\_\_\_\_)$$

Q.30 If ρ_{yx} is negative and high then ratio estimator can not be applied and an alternative estimator is applied and which is known as _________

Q.31 In usual notations, the product estimator $\overline{y}_\rho$ is defined as $\overline{y}_\rho =$ _________

Q.32 If the population mean $\overline{X}_N$ is not known in prior then another estimator is recommended and is defined as $\overline{y}_{pd} =$ _________

Q.33 The product estimator $\overline{y}_\rho$ is a biased estimator of population mean $\overline{Y}_N$, and its amount of bias is expressed as

$$B(\overline{y}_\rho) = \left(\frac{N-n}{nN}\right) \_\_\_\_\_\_\_\_\_$$

Q.34 For large sample sizes, the relation between relations bias and (Cx, Cy) can be expressed as $\frac{B(\overline{y}_\rho)}{\overline{Y}_N} <$ _________

Q.35 For the product estimator $\overline{y}_\rho$ to be more efficient than the sample mean estimator $\overline{y}_n$ the following condition must be satisfied

$$\rho .... \frac{1}{2}\frac{Cx}{Cy}$$

Q.36 If Cx = Cy = C, then the product estimator $\overline{y}_\rho$ will always be more efficient than the sample mean $\overline{y}_n$ for the following condition

ρ ....

Q.37 If the regression line of y *on* x is linear and the line does not pass through the orign, then the most appropriator estimator to estimate $\overline{Y}_N$ is the _________

Q.38 The difference estimator $\overline{y}_D$, to estimate the population mean $\overline{Y}_N$, is defined as $\overline{y}_D = \overline{y}n + k$ (..........) Where k is some constant

Q.39 If the value the constant k is replaced by sample regression coefficient by x then the estimator $\overline{y}_D$ reduces to regression estimator $\overline{y}_{lr}$ defined as $\overline{y}_{lr} = \overline{y}_n +$ _________ (_________)

Q.40 A compact expression for bias $(\overline{y}_{lr})$ can be written as $B(\overline{y}er) =$_________

Q.41 An approximate variance expression for the estimator $\overline{y}_{lr}$ is written as $V(\overline{y}_{lr}) = \frac{N-n}{nN} S_y^{\,2}$ (________)

Q.42 For ρ_{xy} being perfect either positive or negative the sampling variance $V_1(\overline{y}_{lr})$ will be equal to ________

Q.43 The regression estimator $\overline{y}_{lr}$ will always be more efficient than the sample mean $\overline{y}_n$ for the value of ρ_{yx} being not to equal to ________

Q.44 Both the estimators $\overline{y}_{lr}$ and $\overline{y}_n$ will be exactly equally efficeint if ρ is exactly equal to ________

Q.45 If $R_N = \beta_{yx}$ then both the estimators $\overline{y}_{lr}$ and $\overline{y}_R$ will be equally________

Q.46 If the line of regression of y on x is linear but the line does not pass through origin then the following relation holds.

R_N ________ β_{yx}

Q.47 The separate ratio estimator $\overline{y}_{RS}$ is defined as

$$\overline{y}_{RS} = \sum_{i=1}^{n} \text{.......} \overline{y}_{Rh} \text{ for } \overline{y}_{Rh} = \frac{\overline{y}_{nh}}{x_{nh}} \overline{X}_{Nh}$$

Q.48 The expression for $V(\overline{y}_{RS})$ can be written as $V(\overline{y}_{RS}) = \sum_{h}^{K} W_h^2 \left(\frac{N_h - n_h}{n_h - N_h} \right) \left\{ S_h^2 y + \text{..................} \right\}$ in usual notations.

Q.49 For n_h being large (n_h > 30 h) and k is small then bias w. r. to S. E. $(\overline{y}_{RS})$ will be ________

Q.50 The combined ratio estimator $\overline{y}_{RC}$ in stratified random sampling in defined as

$$\overline{y}_{RC} = \text{\_\_\_\_\_\_\_\_} \overline{X}_N \text{ for } \overline{y}_w = \sum_{h=1}^{k} W_h . \overline{y}_{nh}, \ \overline{x}_w = \sum_{h=1}^{n} W_h . \overline{x}_{nh}$$

Q.51 The estimator $\overline{y}_{RC}$ has less risk due to its bias being less than that of the estimator ________

Q.52 When R_h varies from stratum to stratum and $n_h > 30$ V h, than $\overline{y}_{RS}$ is found _________ than $\overline{y}_{RC}$

Q.53 If $n_h < 30, \forall h = 1, 2, k$, then the estimator _________ is preferred to the estimator _________

Q.54 When the value of b_h and b are assumed to be known in prior then both the estimators $\overline{y}_{lrs}$ and $\overline{y}_{lrc}$ will be _________ for the parameter $\overline{Y}_N$.

Q.55 The combined regression estimator $\overline{y}_{lrc}$ is defined as $\overline{y}_{lrc} = \overline{y}_w + -(\overline{X}_N - \overline{x}_w)$

6.3 Multiple Choice Questions and Answers

Q.1 To improve the samling design for the main variable y, an other variable x, highly correlated with y is utilized which is known as.

(a) Auxiliary variable
(b) Anciliary variable
(c) Independent variable
(d) All the three

Q.2 An aggregate value of all the sampling units available of an auxiliary variable x can be uttlized to get an improved estimate of the other variable y which is known as

(a) Study variable
(b) Main variable
(c) Dependent variable
(d) All the three

Q.3 Information available on an auxiliary variable x can be utilized at estimation stage to get an improved estimate of the study variable y through the:

(a) Ratio method of estimation
(b) Product method of estimation
(c) Regression method of estimation
(d) All the three methods mentioned above

Q.4 In general, the most efficient method of estimation to estimate population mean using auxiliary information is-

(a) Ratio method of estimation
(b) Regression method of estimation
(c) Product method of estimation
(d) Sample mean estimator $\overline{y}_n$

Q.5 One of the main aims to use an auxiliary variable x related to study variable y is to -

(a) Get an unbiased estimate of $\overline{y}_N$.

(b) Get an increased precision of $\hat{\overline{y}}_N$.

(c) Get an efficient estimate of $\overline{Y}_N$.

(d) Both (b) and (c)

Q.6 Information on auxiliary variable x can be taken from-

(a) Previous census
(b) Previous records
(c) Current sample
(d) Both (a) and (b)

Q.7 When ρ_{yx} is positive, high and the line of regression of y on x passes through origin then the most reasonable method of estimation in practice is

(a) Regression method
(b) Product method
(b) Ratio method
(d) Sample mean estimator $\overline{y}_n$

Q.8 Let $(\overline{x}_{n,} \overline{y}_n, \overline{X}_N, \overline{Y}_N)$ denote the sample means and population means of the variables (x, y) respectively then ratio estimator $\overline{y}_R$ for $\overline{Y}_N$ is defined as

(a) $\overline{y}_R = \frac{\overline{x}_n}{\overline{y}_n} \overline{X}_N$

(b) $\overline{y}_R = \frac{\overline{y}_n}{\overline{x}_n} \overline{X}_N$

(b) $\overline{y}_R = \frac{\overline{y}_n}{\overline{X}_n} \overline{X}_n$

(b) $\overline{y}_R = \frac{\overline{y}_n}{\overline{x}_n} (N\overline{X}_N)$

Q.9 Let $\overline{x}_{n'}$ be the sample mean of size > n' and $\overline{X}_N$ is unknown then the ratio estimator $\overline{y}_{Rd}$ to estimate $\overline{y}_N$ is defined as-

(a) $\overline{y}_{Rd} = \frac{\overline{y}_n}{x_n} \overline{X}_N$

(b) $\overline{y}_{Rd} = \frac{\overline{y}_n}{x_n} \overline{x}_{n'}$

(c) $\overline{y}_{Rd} = \frac{\overline{y}_n}{\overline{x}_{n'}} \overline{x}_n$

(d) $\overline{y}_{Rd} = \frac{\overline{x}_n}{y_n} \overline{x}_{n'}$

Q.10 The bias of ratio estimator $\overline{y}_R$ is given by the expression

(a) Bias $(\overline{y}_R) = E(\overline{y}_R) - \overline{Y}_N$

(b) Bias $(\overline{y}_R) = -\operatorname{cov}(R_{n,} - \overline{x}_n)$

(c) Bias $(\overline{y}_R) = \dfrac{(N-n)}{nN}\overline{Y}n\,(Cx^2 - \rho Cx\,Cy)$

(d) All the three expressions given above.

Q.11 The expression to work out the I[st] order approximate value of bias $(\overline{y}_R)$ is given as

(a) $B_1(\overline{y}_R) = \dfrac{(N-n)}{nN}\overline{Y}_N\ (Cx^2 - \rho Cx\,Cy)$

(b) $B_1(\overline{y}_R) = \dfrac{(N-n)}{nN}\overline{Y}_N^{\ 2}\ (Cx^2 - \rho Cy\,Cx)$

(c) $B_1(\overline{y}_R) = \dfrac{(N-n)}{nN}\overline{Y}_N(Cx^2 + \rho Cx\,Cy)$

(d) $B_1(\overline{y}_R) = \dfrac{(N-n)}{nN}\overline{Y}_N(Cy^2 + \rho Cx\,Cy)$

Q.12 In case when $Cx = Cy = C$, and ρ_{yx} is perfect positive, $B_1(\overline{y}_R)$ will be

(a) Very high
(b) Very low
(c) Equal to zero
(d) Negative

Q.13 When ρ_{yx} is positive and high then for large sample size n, the bias $(\overline{y}_R)$ will be-

(a) Negative
(b) Very low
(c) Equal to zero
(d) None of these

Q.14 The upper limit of $B(\overline{y}_R)$ is given by

(a) $\sigma_{Rn}\,\sigma_{\overline{x}n}$
(b) $\sigma_{Rn}\,\sigma_{\overline{y}n}$
(c) $\sigma_{RN}\,\sigma_{\overline{x}n}$
(d) None

Q.15 M. S. E $(\bar{y}_R)$ is defined by the expression

(a) $E(\bar{y}_R - \bar{Y}_N)^2$ (b) $E(\bar{y}_R - E\,\bar{Y}_R)^2$

(c) $E(E(\bar{y}_R) - \bar{Y}_N)^2$ (d) $E\,(\bar{y}_R - E(\bar{y}_n^2))$

Q.16 The first order approximation to M. S. E. $(\bar{y}_R)$ can be obtained from the expression given as

(a) $M_1\ (\bar{y}_R) = \left(\dfrac{N-n}{n\,N}\right) \bar{Y}_N{}^2(Cy^2 + R^2{}_N Cx^2 - 2\rho\, Cx\, Cy)$

(b) $M_1\ (\bar{y}_R) = \left(\dfrac{N-n}{n\,N}\right) \bar{Y}_N{}^2(Cy^2 + R^2{}_N Cx^2 + 2\rho\, Cx\, Cy)$

(c) $M_1\ (\bar{y}_R) = \left(\dfrac{N-n}{n\,N}\right) R_N{}^2(Cy^2 + R^2{}_N Cx^2 - 2\rho\, Cx\, Cy)$

(d) $M_1\ (\bar{y}_R) = \left(\dfrac{N-n}{n\,N}\right) \bar{Y}_N{}^2(Cy^2 + R^2{}_N Cx^2 - 2R_N\, Cx\, \dot{C}y)$

Q.17 In terms of Sy^2, Sx^2 and R_N, the first order approximation to the M.S.E. $(\bar{y}_R)$ can also be expressed as-

(a) $M_1\ (\bar{y}_R) = \left(\dfrac{N-n}{n\,N}\right) (Sy^2 + R_N{}^2\, Sx^2 - 2R_N\, Sxy)$

(b) $M_1\ (\bar{y}_R) = \left(\dfrac{N-n}{n\,N}\right) (Sy^2 + R_N{}^2\, Sx^2 + 2R_N\, Sxy)$

(c) $M_1\ (\bar{y}_R) = \left(\dfrac{N-n}{n\,N}\right) \bar{Y}_N^2\, (S_y^2 + R_N^2 S_x^2 - 2R_N\, Sxy)$

(d) $M_1\ (\bar{y}_R) = \left(\dfrac{N-n}{n\,N}\right) \left(S_y{}^2 + R_N{}^2 + S_x^2 - 2R_N\, \rho\, S_x S_y\right)$

Q.18 The first order approximation to the bias of the estimator can be written as-

(a) $B_1\ (\overline{y}_{Rd}) = \left(\frac{n' - n}{n\,n'}\right)\ \overline{Y}_N (Cx^2 + \rho Cx\, Cy)$

(b) $B_1 (\overline{y}_R) = \left(\frac{n' - n}{n\,n'}\right)\ \overline{Y}_N (Cx^2 - \rho Cx\, Cy)$

(c) $B_1 (\overline{y}_{Rd}) = \left(\frac{N - n'}{n' N}\right)\ \overline{Y}_N (Cx^2 + \rho Cx\, Cy)$

(d) $B_1 (\overline{y}_{Rd}) = \left(\frac{N - n'}{n' N}\right)\ \overline{Y}_N (Cx^2 - \rho Cx\, Cy)$

Q.19 Under certain conditions the ratio estimate $\overline{y}_R$ is found more efficient than the sample mean $\overline{y}_n$ if-

(a) $\rho_{yx} \geq \frac{1}{2}\frac{cx}{cy}$ (b) $\rho_{yx} > \frac{1}{2}\frac{cx}{cy}$

(c) $\rho_{yx} > \frac{1}{2}\frac{cy}{cx}$ (d) $\rho_{yx} < \frac{1}{2}\frac{cx}{cy}$

Q.20 Under large sample and $Cx = Cy$ the ratio estimator $\overline{y}_R$ is found more efficient than the sample mean $\overline{y}_n$ if

(a) $\rho \geq 0.5$ (b) $\rho > 0.5$

(c) $\rho < 0.5$ (d) $\rho \leq 0.5$

Q.21 An estimate of $B_1 (\overline{y}_R)$ can be obtained from an expression written as

(a) $\hat{B}_1 (\overline{y}_R) = \left(\frac{N - n}{n\,N}\right) \overline{Y}_N^{\ 2} (Cx^2 - \rho Cx\, Cy)$

(b) $\hat{B}_1 (\overline{y}_R) = \left(\frac{N - n}{n\,N}\right) \overline{Y}_n (Cx^2 - r\, c_x\, c_y)$

(c) $\hat{B}_1 (\overline{y}_R) = \left(\frac{N - n}{n\,N}\right) \overline{y}_n^{\ 2} (c^2_x + r\, c_x\, c_y)$

(d) $\hat{B}_1 (\overline{y}_R) = \frac{N - n}{n\,N} \overline{y}_n^{\ 2} (Cx^2 - r\, Cx\, Cy)$

where $\overline{Y}_{n,}\ Cx, Cy,\ r$ are the sample value.

Q.22 $M_1(\bar{y}_R)$ can be estimated through an expression written as

(a) $\hat{M},(\bar{y}_R) = \left(\frac{N-n}{nN}\right)(S^2{}_y + R_n^2\, s^2 x - 2Rn\, sxy)$

(b) $\hat{M},(\bar{y}_R) = \left(\frac{N-n}{nN}\right)(s^2{}_y + R_n^2\, s^2 x + 2Rn\, sxy)$

(c) $\hat{M}_1(\bar{y}_R) = \left(\frac{N-n}{nN}\right)\bar{y}^2{}_n\,(c_y^2 + c_x^2 - 2r\, cx\, cy)$

(d) Both (a) and (c)

Q.23 $B_1(\bar{y}_{Rd})$ can be estimated through an expression written as

(a) $\hat{B}_1(\bar{y}_{Rd}) = \left(\frac{N-n'}{n'N}\right)\bar{y}_n\,(Cx^2 - r\,Cx\,Cy)$

(b) $\hat{B}_1(\bar{y}_{Rd}) = \left(\frac{N-n'}{n\,n'}\right)\bar{y}_n\,(cx^2 - r\,cx\,cy)$

(c) $\hat{B}_1(\bar{y}_{Rd}) = \left(\frac{n'-n}{nn_1}\right)\bar{y}_n{}^2\,(C_x^2 - r\,Cx\,Cy)$

(d) $\hat{B}_1(\bar{y}_{Rd}) = \bar{y}_n\,(C_x^2 - r\,Cx\,Cy)$

Q.24 An unbiased ratio type estimator $\bar{y}_{Ru}$ suggested by Hartley and Ross (1954) is defined as

(a) $\bar{y}_{Ru} = \bar{r}_n\,\bar{x}_n + \frac{n(N-1)}{(n-1)(N)}(\bar{y}_n - \bar{r}_n\,\bar{x}_n)$

(b) $\bar{y}_{Ru} = \bar{r}_n\,\bar{y}_n + \frac{n(N-1)}{(n-1)N}(\bar{y}_n - \bar{r}_n\,\bar{y}_n)$

(c) $\bar{y}_{Ru} = \bar{r}_n\,\bar{y}_n + \frac{n(N-1)}{(n-1)N}(\bar{y}_n + \bar{r}_n\,\bar{x}_n)$

(d) $\bar{y}_{Ru} = \bar{r}_n\,\bar{x}_n + \frac{n(N-1)}{(n-1)N}(\bar{y}_n + \bar{r}_n\,\bar{x}_n)$

where $\bar{r}_n = \frac{1}{n}\sum_{i}^{n} ri \quad r_i = (y_i / x_i, i = 1, 2..........n)$

Q.25 The sampling variance of the estimator $\overline{y}_{RU}$ can be worked out using the expression

(a) $V_1(\overline{y}_{RU}) = \frac{1}{n}(S_y^{\ 2} + \overline{r}_N^{\ 2} S_x^{\ 2} - 2\overline{r}_N Sxy)$

(b) $V_1(\overline{y}_{RU}) = \frac{1}{n}(S_y^{\ 2} + \overline{r}_N^{\ 2} S_x^{\ 2} + 2\overline{r}_N Sxy)$

(c) $V_1(\overline{y}_{RU}) = \frac{1}{N}(S_y^{\ 2} + \overline{r}_N^{\ 2} S_x^{\ 2} - 2\overline{r}_N Sxy)$

(d) $V_1(\overline{y}_{RU}) = \frac{1}{N}(S_y^{\ 2} + \overline{r}_N^{\ 2} S_x^{\ 2} + 2\overline{r}_N Sxy)$

Q.26 When ρ_{yx} is negative and high then most appropriate method of estimation to estimate the population mean $\overline{Y}_N$ using auxilliary information on x is

(a) Ratio method (b) Product method

(c) Regression method (d) None of the above

Q.27 If the value of population mean $\overline{X}_N$ is not known in prior then other product estimator $\overline{y}_{pd}$ is recommended and is defined as

(a) $\overline{y}_{pd} = \frac{\overline{y}_n \overline{x}_{n'}}{\overline{x}_n}$ (b) $\overline{y}_{pd} = \frac{\overline{y}_n \overline{x}_n}{\overline{x}_{n'}}$

(c) $\overline{y}_{pd} = \frac{\overline{y}_n \overline{x}_{n'}}{\overline{X}_N}$ (b) None of these

where $\overline{x}_{n'}$ is a sample mean of size n' > n.

Q.28 The product estimator $\overline{y}_p$ is a biased estimator and its amount of bias is expressed as

(a) $B(\overline{y}_p) = \left(\frac{N-n}{n\,N}\right)\frac{sxy}{\overline{X}_N}$ (b) $B(\overline{y}_p) = \left(\frac{N-n}{n\,N}\right)\frac{Sxy}{\overline{X}_N}$

(c) $B(\overline{y}_p) = \left(\frac{N-n}{n\,N}\right)\frac{Sxy}{\overline{Y}_N}$ (d) $B(\overline{y}_p) = \left(\frac{N-n}{n\,N}\right)\frac{Sxy}{\overline{X}_N\overline{Y}_N}$

Q.29 For large sample approximation the relation between relative bias and (Cx, Cy) exists as

(a) $\frac{B(\overline{y}_p)}{\overline{Y}_N} < Cx\,Cy$ (b) $\frac{B(\overline{y}_p)}{\overline{Y}_N} > Cx\,Cy$

(c) $\frac{B(\overline{y}_p)}{\overline{Y}_N} \geq Cx\,Cy$ (d) $\frac{B(\overline{y}_p)}{\overline{X}_N} \geq Cx\,Cy$

Q.30 In case of large sample approximaton, an expression for $V_1(\overline{y}_p)$ to the first order approximation is written as

(a) $V_1(\overline{y}_p) = \frac{N-n}{n\,N}\left(Sy^2 + R_N{}^2\,Sx^2 - 2R_N Sxy\right)$

(b) $V_1(\overline{y}_p) = \left(\frac{N-n}{n\,N}\right)\left(Sy^2 + R_N{}^2\,Sx^2 + 2R_N Sxy\right)$

(c) $V_1(\overline{y}_p) = \frac{N-n}{n\,N}\left(Sy^2 + R_N{}^2\,Sx^2 + 2R_N Sx\,Sy\right)$

(d) $V_1(\overline{y}_p) = \frac{N-n}{n\,N}\left(Sy^2 + R_N{}^2\,Sx^2 - 2R_N Sx\,Sy\right)$

Q.31 Bias $(\overline{y}_p)$ can be estimated through an expression given as

(a) $\hat{B}_1(\overline{y}_p) = \frac{N-n}{n\,N}\,\frac{Sxy}{\overline{x}_n}$ (b) $\hat{B}_1(\overline{y}_p) = \frac{N-n}{n\,N}\,\frac{sxy}{\overline{x}_n}$

(c) $\hat{B}_1(\overline{y}_p) = \frac{N-n}{n\,N}\,\frac{Sxy}{\overline{X}_N}$ (d) $\hat{B}_1(\overline{y}_p) = \frac{N-n}{n\,N}\,\frac{sx\,sy}{\overline{x}_n}$

Q.32 The sampling variance $V_1(\overline{y}_p)$ can be estimated through an expression written as -

(a) $\hat{V}_1(\overline{y}_p) = \frac{(N-n)}{n\,N}\,(sy^2 + Rn^2\,sx^2 - 2Rn\,sxy)$

(b) $\hat{V}_1(\overline{y}_p) = \frac{(N-n)}{n\,N}\,(sy^2 + Rn^2\,sx^2 + 2Rn\,sxy)$

(c) $\hat{V}_1(\overline{y}_p) = \dfrac{N-n}{nN}(Sy^2 + R_n^2 S_x^2 + 2Rn\,sxy)$

(d) $\hat{V}_1(\overline{y}_p) = \dfrac{N-n}{nN}(sy^2 + Rn^2\,sx^2 + 2Rn\,Sxy)$

Q.33 The the product estimator $\overline{y}_p$ to be more efficient than the sample mean $\overline{y}_n$ the condition is

(a) $\rho < \dfrac{1}{2}\dfrac{Cx}{Cy}$ (b) $\rho < \dfrac{1}{2}\dfrac{Cy}{Cx}$

(c) $\rho > \dfrac{1}{2}\dfrac{Cx}{Cy}$ (d) $\rho > \dfrac{1}{2}\dfrac{Cy}{Cx}$

Q.34 If the line of y on x is linear but the line does not pass through then most suitable method of estimation is

(a) Ratio method (b) Regression method

(c) Product method (d) Both (a) and (c)

Q.35 The difference estimator $\overline{y}_D = \overline{y}_n + k(\overline{X}_N - \overline{x}_n)$ is analogus to -

(a) Ratio estimator (b) Procuct estimator

(c) Regression estimator (d) Ratio cum product estimator

Q.36 The regression estimator $\overline{y}_{lr}$ to estimate the population mean $\overline{y}_N$ is defined as

(a) $\overline{y}_{lr} = \overline{y}_n + bxy(\overline{X}_N - \overline{x}_n)$ (b) $\overline{y}_{lr} = \overline{y}_n - byx(\overline{X}_N - \overline{x}_n)$

(c) $\overline{y}_{lr} = \overline{y}_n + byx(\overline{x}_n - \overline{X}_N)$ (d) $\overline{y}_{lr} = \overline{y}_n + byx(\overline{X}_N - \overline{x}_n)$

Q.37 An approximate compact expression for bias $(\overline{y}_{pr})$ can be written as

(a) $B_1(\overline{y}_{lr})$ = covariance $(byx, \overline{x}_n)$

(b) $B_1(\overline{y}_{lr})$ = - covariance $(byx, \overline{x}_n)$

(c) $B_1(\overline{y}_{lr})$ = - covariance $(\overline{x}n, byx)$

(d) $B_1(\overline{y}_{lr})$ = - Both (b) and (c)

Q.38 An approximate variance expression for the estimator $\overline{y}_{lr}$ is written as.

(a) $V_1(\overline{y}_{lr}) = \frac{N-n}{nN} S_y^2 (1 + \rho_{yx}^2)$

(b) $V_1(\overline{y}_{lr}) = \frac{N-n}{nN} S_y^2 (1 - \rho_{yx}^2)$

(c) $V_1(\overline{y}_{lr}) = \frac{N-n}{nN} S_y^2 (1 - \rho_{yx})$

(d) $V_1(\overline{y}_{er}) = \frac{N-n}{nN} Sy^2 (1 + \rho yx)$

Q.39 In case of correlation coefficient ρ_{xy} being high (either possitive or negative the following relation is true.)

(a) $V_1(\overline{y}_{lr}) > V(\overline{y}_n)$ (b) $V_1(\overline{y}_{lr}) = V(\overline{y}_n)$

(c) $V_1(\overline{y}_{lr}) =< V(\overline{y}_n)$ (d) None of the above

Q.40 In case of ρ_{yx} being perfect positive or perfact negative __________ the following relation holds-

(a) $V_1(\overline{y}_{lr}) = V(\overline{y}_n)$ (b) $V(\overline{y}_{lr}) > V(\overline{y}_n)$

(c) $V(\overline{y}_{lr}) > 0.0$ (d) $V(\overline{y}_{lr}) = 0.0$

Q.41 An expression to estimate $V_1(\overline{y}_{lr})$ can be written as

(a) $\hat{V}_1(\overline{y}_{lr}) = \frac{N-n}{nN} Sy^2 (1 - r^2 xy)$

(b) $\hat{V}_1(\overline{y}_{lr}) = \frac{N-n}{nN} sy^2 (1 - r^2 xy)$

(c) $\hat{V}_1(\overline{y}_{lr}) = \frac{N-n}{nN} s_y^2 (1 - r_{yx}^2)$

(d) Both (b) and (c)

Q.42 If ρ_{yx} is not equl to zero then the following relation holds good.

(a) $V_1(\overline{y}_{lr}) = V(\overline{y}_n)$ (b) $V_1(\overline{y}_{lr}) < V(\overline{y}_n)$

(b) $V_1(\overline{y}_{lr}) \geq V(\overline{y}_n)$ (d) $V(\overline{y}_n) \leq V(\overline{y}_{lr})$

Q.43 If the value of ρ_{xy} is exactly equal to zero then the following result is true.

(a) $V_1(\overline{y}_{lr}) < V(\overline{y}_n)$ (b) $V_1(\overline{y}_{lr}) = V(\overline{y}_n)$

(c) $V_1(\overline{y}_{lr}) \neq V(\overline{y}_n)$ (d) $V(\overline{y}_n) > V(\overline{y}_{lr})$

Q.44 If $R_N = \beta_{yx}$ than the following result is true.

(a) $V(\overline{y}_R) > V(\overline{y}_{lr})$ (b) $V(\overline{y}_R) < V(\overline{y}_{lr})$

(c) $V(\overline{y}_R) = V(\overline{y}_{lr})$ (d) $V(\overline{y}_{lr}) \neq V(\overline{y}_R)$

Q.45 If the regression line of y on x is linear and the line passes through origin then the following relation holds good

(a) $R_N > \beta_{yx}$ (b) $R_N = \beta_{yx}$

(c) $R_N < \beta_{yx}$ (d) $R_N \neq \beta_{yx}$

Q.46 In case of stratified random sampling with k strata with other usual notations the separate ratio estimator is defined for $\overline{Y}_N$ as

(a) $\overline{y}_{RS} = \sum_{h=1}^{k} W_h \overline{y}_{Rh}$ (b) $\overline{y}_{RS} = \frac{1}{k}\sum_{i=1}^{k} W_h . \overline{y}_{Rh}$

(c) $\overline{y}_{RS} = \frac{1}{N}\sum_{i}^{k} W_h \overline{y}_{Rh}$ (d) $\overline{y}_{RS} = \sum_{i}^{k} \frac{W_h}{n_h} \overline{y}_{Rh}$

Q.47 The combined ratio estimator in stratified random sampling is defined as

(a) $\overline{y}_{RC} = \frac{\overline{y}_w}{\overline{x}_w} \overline{X}_N$ (b) $\overline{y}_{RC} = \frac{\overline{y}_w}{\overline{x}_w} \overline{x}_n$

(c) $\overline{y}_{RC} = \frac{\overline{x}_w}{\overline{y}_w} \overline{X}_N$ (d) None of the above

Where $\overline{y}_w = \sum_{h=1}^{k} W_h \overline{y}_{nh}$, $\overline{x}_w = \sum_{i}^{k} W_h \overline{x}_{nh}$

Q.48 If R_h varies from stratum to stratum and $n_h > 30$ for all h = 1, 2, ...,k then following result holds good.

(a) $V(\overline{y}_{RS}) > V(\overline{y}_{RC})$ (b) $V(\overline{y}_{RS}) < V(\overline{y}_{RC})$

(c) $V(\overline{y}_{RC}) > V(\overline{y}_{RS})$ (d) Both (b) and (c)

Q.49 If ($\overline{y}_w$ $\overline{x}_w$. by$x$) are the weighted sample means of (y, x)and sample regression coefficient by$x$ same for all strata then combined regression estimator $\overline{y}_{lrc}$ is defined as-

(a) $\overline{y}_{lrc} = \overline{y}_w + b_{yx}(\overline{x}_N - \overline{x}_w)$

(b) $\overline{y}_{lrc} = \overline{y}_W + b_{yx}(\overline{x}_W - \overline{x}_N)$

(c) $\overline{y}_{lrc} = \overline{y}_W + b_{xy}(\overline{X}_N - \overline{x}_w)$

(d) $\overline{y}_{lrc} = \overline{y}_W - b_{xy}(\overline{x}_w - X_N)$

Q.50 When p_h is assumed to be same for all strata then

(a) $\overline{y}_{\ell rc}$ is recommended in practice

(b) $\overline{y}_{\ell rs}$ is recommended in practice

(c) $V(\overline{y}_{\ell rs}) < V(\overline{y}_{lrs})$

(d) Both (a) and (c)

Q.51 When $n_h < 30$ for all strata the following result is true.

(a) $B(\overline{y}_{RS}) > B(\overline{y}_{RC})$

(b) $B(\overline{y}_{RS}) < B(\overline{y}_{RC})$

(c) $B(\overline{y}_{RS}) = B(\overline{y}_{RC})$

(d) None of the above

6.4 Key Answers

6.4.1 Frue / False

1.	True	2.	True	3.	True
4.	False	5.	False	6.	True
7.	True	8.	True	9.	False
10.	True	11.	True	12.	False
13.	False	14.	False	15.	True
16.	False	17.	True	18.	False
19.	False	20.	True	21.	False
22.	True	23.	False	24.	True
25.	True	26.	True	27.	True
28.	False	29.	Flase	30.	False
31.	False	32.	False	33.	False
34.	False	35.	False	36.	True
37.	False	38.	True	39.	False

40. True	41. True	42. True
43. True	44. True	45. True
46. True	47. True	48. True
49. True	50. True	51. True
52. True	53. False	54. True
55. False	56. False	57. True
58. True	59. True	60. False
61. False	62. True	63. True
64. True	65. True	66. True
67. True	68. False	69. True
70. True	71. False	72. False
73. False	74. True	75. False
76. True	77. True	78. True
79. True	80. True	81. True
82. True	83. True	84. True
85. True		

6.4.2 Fill in the Blank

1. Auxiliary variable
2. Auxiliary variable
3. Auxiliary
4. Auxiliary
5. Precision
6. Independent or covariate
7. Dependent, Main
8. Census
9. Ratio Method
10. $\frac{\overline{y}_n}{\overline{X}_n} \overline{X}_N$
11. $\frac{\overline{y}_n}{\overline{x}n} \overline{x}_{n1}$
12. $\overline{y}_n / \overline{x}_n$

13. $E(\overline{y}_R) - \overline{Y}_N$

14. $-\text{Cov}(Rn, \overline{x}_n)$

15. $-\rho Cx\, Cy$

16. $C^2(1-\rho)$

17. Positive high

18. Zero

19. Zero

20. $\sigma_{Rn}\sigma_{\overline{x}n}$

21. $\sigma_{Rn}\sigma_{\overline{x}n}$

22. $E(\overline{y}_R - \overline{y}_N)^2$

23. $E(\overline{y}_R - E(\overline{y}_R))^2$

24. $(S^2{}_y + R^2{}_N S^2 x - 2R_N Sxy)$

25. $(C^2{}_y + C^2{}_x - 2\rho\, Cx\, Cy)$

26. $\rho\, Cx\, Cy$

27. $\frac{1}{2}\frac{Cx}{Cy}$

28. $\rho > \frac{1}{2}$

29. $(C_x^2 + rCx\, Cy)$

30. Product estimator

31. $= \frac{\overline{y}_n\, \overline{x}_n}{\overline{X}_N}$

32. $= \frac{\overline{y}_n\, \overline{x}_n}{\overline{X}_{n1}}$

33. $= \overline{Y}_N \frac{Sxy}{\overline{X}_N}$

34. $< Cx\,Cy$
35. $<$
36. $< \frac{1}{2}$
37. Regression estimator
38. $(\overline{X}_N - \overline{x}n)$
39. $byx\,(\overline{X}_N - \overline{x}_n)$
40. $-\operatorname{cov}(by_x\,\overline{x}_n)$
41. $(1-\rho^2)$
42. Zero
43. Zero
44. Zero
45. Efficient
46. $\neq$
47. W_h
48. $(R^2{}_{Nh}S^2{}_{hx} - 2R_{Nh}S_{hxy})$
49. Negligble
50. $\overline{y}_w\ /\ \overline{x}_n$
51. $\overline{y}_{RS}$
52. More efficient
53. $\overline{y}_{RC}, \overline{y}_{RS}$
54. Unbiased
55. byx

6.4.3 Multiple Choice Questions and Answers

1.	d	2.	d	3.	d
4.	b	5.	d	6.	d
7.	c	8.	b	9.	b
10.	d	11.	a	12.	c
13.	a	14.	a	15.	a
16.	a	17.	a	18.	a
19.	b	20.	b	21.	b
22.	d	23.	b	24.	a
25.	a	26.	b	27.	b
28.	b	29.	a	30.	b
31.	b	32.	b	33.	a
34.	b	35.	c	36.	d
37.	d	38.	b	39.	c
40.	d	41.	d	42.	b
43.	b	44.	c	45.	b
46.	a	47.	a	48.	d
49.	a	50	b	51.	a

7

Multistage Sampling

7.1 True / False

Q.1 Cluster sampling restricts the spread of sample over the population which results generally in increased sampling variance of estimators under considerataion.

Q.2 In adopting cluster sampling, few selected clusters are enumerated partially.

Q.3 By increasing the number of clusters and decreasing the cluster sizes, the spread of sample over the population can be increased.

Q.4 By increasing the number of clusters and by decreasing the cluster size, we get more representative sample.

Q.5 If all the selected clusters are enumerated partially then such sampling procedure adopted, is known as "sub-sampling or two stage sampling."

Q.6 In sub sampling if the cluster are treated as units of selection at first stage then such clusters are known as first stage units (F.S.U.) or primary stage units.

Q.7 A sampling procedure which consists of three or more than three stages for sample selection, is known as multistage sampling scheme.

Q.8 In multistage sampling scheme the different sampling schemes can not be adopted independently at different stages of sample selection.

Q.9 Multistage sampling procedure has been found very useful in practice and this procedure is commonly adopted in large sample surveys.

Q.10 Multistage sampling procedure can be expected less efficient than single stage sampling

Q.11 From the preparation of sampling frame point of view, the multistage sampling procedure is advantages.

7.2 Notations for Two Stage Sampling with Equal F. S. U

Let us define

N : The number of clusters (F. S. U) in the population

M : The size of each clusters

n : The number of clusters selected in the sample

NM : The number of ultimate stage units in the population.

y_{ij} : jth S.S.U in the ith F.S.U

$\overline{y}_i$: $\frac{1}{\overline{n}}\sum_{U=1}^{n} y_{ij}$: Sample mean of the sample drawn from ith F. S.U.

$\overline{y}_{nm}$: $\frac{1}{n}\sum_{i}^{n}\overline{y}_i = \frac{1}{n}\frac{1}{m}\sum_{i=1}^{n}\sum_{j=1}^{m} y_{ij}$ = sample mean

Q.12 The sample mean $\overline{y}_{nm}$ is a biased estimate of population mean $\overline{Y}_N$

Q.13 The sampling variance of the sample mean $\overline{y}_{nm}$ can be worked out from the expression written as

$$V(\overline{y}_{nm}) = \frac{N-n}{nN} S_b^{\ 2} + \frac{1}{n}\frac{(M-n)}{nM}\overline{S}_w^{\ 2}$$

Q.14 The between cluster mean square $S_b^{\ 2}$ and within cluster mean square $\overline{S}_w^{\ 2}$ are defined as

$$S_b^{\ 2} = \frac{1}{N-1}\sum_{i}^{N}(\overline{Y}_i - \overline{Y})^2 \text{ and}$$

$$\overline{S}_w^{\ 2} = \frac{1}{N}\sum_{i}^{N} S_i^2 = \frac{1}{N}\sum_{i}^{N}\sum_{j=1}^{M}\frac{\left(y_{ij} - \overline{y}_i\right)^2}{(M-1)}$$

Q.15 For sufficiently large value of N and M the quantities $\left(\frac{N-n}{N}\right)$ and $\left(\frac{M-m}{M}\right)$ are small and can be ignored then $V(\overline{y}_{nm})$ reduces to the form $V(\overline{y}_{nm}) = \frac{S_b^{\ 2}}{n} + \frac{\overline{S}_w^{\ 2}}{n\,m}$

Q.16 The second term $\frac{1}{n}\left(\frac{M-m}{mM}\right)\overline{S}_w{}^2$ in $V(\overline{y}_{nm})$ represents the contribution to the sampling variance of $\overline{y}_{nm}$ arising due to sub-sampling done from n selected first stage units.

Q.17 In case of sub sampling if n = N then sub sampling reduces to stratified random sample each of size m drawn from each N strata.

Q.18 Cluster sampling and stratified random sampling schemes can not be considered as particular cases of two stage sampling under certain conditions.

Q.19 The variance cannot get estimated unbiasedly using the following expression.

$$\hat{V}(\overline{y}_{nm})=\left(\frac{N-n}{nN}\right)s_b{}^2+\frac{1}{N}\left(\frac{M-m}{mM}\right)s_w{}^2$$

Where $s_b{}^2 = \left(\frac{1}{n-1}\right)\sum_{i=1}^{n}(\overline{y}_i-\overline{y}_{nm})^2$ and $s_w{}^2=\frac{1}{n}\sum_{i}^{n}s_i{}^2$

$$s_i{}^2=\frac{1}{(m\text{-}1)}\sum_{i=1}^{m}(y_{ij}-\overline{y}_i)^2$$

7.3 Two Stage Sampling with Unequal S.S.U. & F.S.U.

Let : Mi : Size of ith f.s.u.

m_i : Size of ith sample drawn from ith selected f.s.u.

$$\overline{y}_{imi} = \frac{1}{m_i}\sum_{j=1}^{mi}y_{ij}$$

$$\overline{M} = \frac{1}{N}\sum_{i=1}^{N}M_i$$

Q.20 An unbiased estimator of the population mean $\overline{Y}_N$ is defined as

$$\overline{y}_u=\frac{1}{n}\sum_{i}^{n}\left(\frac{M_i}{\overline{M}}\right)\overline{y}_{imi}$$

Q.21 In case of three stage sampling procedure with equal number of units at all three stages in population and samples written as (N, M, P) & (n, m, p) respectively, an unbiased estimator for population mean $\overline{y}$... cannot be defined as

$$\overline{y}_{nmp} = \frac{1}{n}\frac{1}{m}\frac{1}{p}\sum_{i=1}^{n}\sum_{j=1}^{m}\sum_{i=j=1}^{p} y_{ijk}$$

Q.22 The terms "Two phase sampling" and "Double sampling" do not denote the same sampling scheme.

Q.23 Mulitiphase sampling and Multistage sampling are not two different sampling scheme.

Q.24 In multiphase sampling the sampling units are same in all the phases where as in multistage sampling, the sampling units differ in sizes at different stages.

Q.25 In multistage sampling the size of sampling units at any stage will be larger than the size of sampling unit at its succeeding stage.

7.4 Fill in the Blanks

Q.1 Cluster sampling restricts the spread of sample over the population which results generally in increased ____________ of estimators under consternation.

Q.2 In adopting cluster sampling, few selected clusters are ____________ completely.

Q.3 By increasing the number of clusters and decreasing cluster sizes, the spread of samples over the ____________ can be increased.

Q.4 By increasing the number of clusters and decreasing cluster sizes, the spread of sample over the population can be ____________

Q.5 By increasing the number of clusters and decreasing cluster sizes, we get more ____________ sample.

Q.6 If all the selected clusters are enumerated partially then such sampling procedure adopted is known as ____________" or "____________"

Q.7 In sub sampling, if the cluster are treated as units of selection at first stage then such clusters are known as "____________ unit" or "____________ units."

Q.8 A sampling procedure which consists of three or more than three stages for sample selection is known as "____________" sampling.

Q.9 In multistage sampling scheme, different ____________ can be adopted ____________ at different stages of sample selection

Q.10 Multistage sampling scheme has been found more useful in practice and this scheme is commonly adopted in ____________ survey.

Q.11 Multistage sampling is considered as a better combination of sampling schemes if it is the combination of ____________ sampling and ____________ sampling.

Q.12 Multistage sampling prodedure is considered (expected) less efficient than ____________ sampling.

Q.13 The sample mean $\overline{y}_{nm}$ is an (____________) estimator of population mean $\overline{Y}_N$.

Q.14 The sampling variance of the sample estimator $\overline{y}_{nm}$ can be worked out from the expression written as

$$V(\overline{y}_{nm}) = \left(\frac{N-n}{nN}\right) \ldots\ldots\ldots\ldots + \frac{1}{n}\left(\frac{M-m}{mM}\right) \ldots\ldots\ldots\ldots$$

Q.15 Between cluster mean square $S_b^{\ 2}$ in defined as

$$S_b^{\ 2} = \frac{1}{(N-1)} \sum_{i}^{N} (\ldots\ldots\ldots\ldots)^2$$

Q.16 Within cluster mean square $\overline{S}_w^{\ 2}$ is defined as $(Mi/\overline{M})$

Where $$\overline{S}_i^{\ 2} = \frac{1}{(\ldots\ldots.)} \sum_{i=1}^{M} (y_i - \overline{y}_i)^2$$

Q.17 For sufficiently large value of N and M th quantities $\frac{(N-n)}{N}$ and $\frac{(M-m)}{M}$ can be approximated by 1, then $V(\overline{y}_{nm})$ reduces to the form

$$V(\overline{y}_{nm}) \approx \frac{S_b^{\ 2}}{\ldots\ldots.} + \frac{\overline{S}_w^{\ 2}}{\ldots\ldots.}$$

Q.18 The second term $\frac{1}{n}\left(\frac{M-n}{mN}\right)\overline{S}_w^{\ 2}$ in $V(\overline{y}_{nm})$ represents the contribution to the arising due to ____________ done from n selected first stage units.

Q.19 In case of sub-sampling if n = N then sub-sampling scheme reduces to ____________

Q.20 Cluster sampling can be regarded as particular case of ____________ under certain condition.

Q.21 ____________ stratified random sampling can be regarded as a particular case of ____________ when n = N.

Q.22 The variance $V(\overline{y}_{nm})$ can be estimated unbiasedly with the help of following expression-

$$\hat{V}(\overline{y}_{nm}) = \left(\frac{N-n}{nN}\right).....+\frac{1}{N}\left(\frac{M-m}{mM}\right).....$$

Q.23 In case of unequal F. S. U and S. S. U an unbiased estimate of population mean $\overline{y}_N$ can be defined as

$$\hat{V}(\overline{y}_u) = \frac{1}{n}\sum_{i}^{n}(....)\overline{y}_{imi}$$

Q.24 The term "Two phase sampling" and "Double sampling denote the ____________ sampling scheme.

Q.25 Multiphase and multistage sampling schemes are two ____________ types of sampling schemes.

Q.26 In multiphase sampling scheme, the sampling units are the ____________ in all the phases.

Q.27 In multistage sampling, the size of sampling units are ____________ at differenat stages of sampling.

Q.28 In multistage sampling, the size of sampling units at any stage will be ____________ than the size of sampling units at its succeeding stages.

7.5 Multiple Choice Questions and Answers

Q.1 Due to restricton imposed on cluster sampling, the spread of sample over population, the sampling variance of sample means is generally-

(a) Decreased (b) Increased

(c) Remains same (d) None of the three

Q.2 In adopting two stage sampling, few selected F.S.U. are enumerated-

(a) Partially

(b) Completely

(c) Systematically

(d) Nothing can be said with surity

Q.3 If all the selected clusters are enumerated partially then such sampling scheme adopted is known as-

(a) Sub sampling (b) Partial sampling

(c) Two stage sampling (d) Both (a) and (c)

Q.4 In sub sampling, if clusters are treated as sampling units at first stage then clusters are known as-

(a) Primary stage unit (b) Ultimate stage unit

(c) First stage unit (d) Both (a) & (c)

Q.5 A sampling scheme which consists of three or more stages for ultimate selection, is known as

(a) Multiphase sampling (b) Multistage sampling

(c) Double sampling (d) Both (a) and (b)

Q.6 In multistage sampling scheme different sampling schemes can be adopted at different stages as

(a) In independent way (b) In dependent way

(c) In preceeding way (d) In succeeding way

Q.7 Multistage sampling scheme is commonly used in-

(a) Large sample survey (b) Small sample survey

(c) In complete enumeration (d) Multipurpose survey

Q.8 Multistage sampling scheme is considered as a better combination of sampling schemes if it is the combination of-

(a) Random sampling and systematic sampling

(b) Random sampling and cluster sampling

(c) Random sampling and stratified random sampling

(d) Cluster sampling and purposive sampling

Q.9 Multistage sampling as compared with single stage sampling is expected to be-

(a) Less efficient (b) More efficient
(c) Equally efficient (d) None of the three

Q.10 From the preparation of sampling frame point of view, the multistage sampling procedure is

(a) Advantages (b) Disadvantageous
(c) Costly (d) Complicated

Q.11 In two stage sampling, the sample mean estimator $\overline{y}_{nm}$ estimates the population mean $\overline{Y}_N$

(a) Biasedly
(b) Unbiasedly
(c) Consistently
(d) Less efficiently than sample median

Q.12 The expression for $V(\overline{y}_{nm})$ given below is correct

(a) $V(\overline{y}_{nm}) = \frac{N-n}{nN} S_b^2 + \frac{1}{N}\frac{(M-m)}{mN}\overline{S}_w^2$

(b) $V(\overline{y}_{nm}) = \frac{N-n}{nN} S_b^2 + \frac{1}{n}\frac{(M-m)}{mM}\overline{S}_w^2$

(c) $V(\overline{y}_{nm}) = \frac{N-n}{nN} \overline{S}_w^2 + \frac{1}{n}\frac{(M-m)}{mM}\overline{S}_b^2$

(d) $V(\overline{y}_{nm}) = \frac{N-n}{nM} \overline{S}_w^2 + \frac{1}{n}\frac{(M-m)}{nN}\overline{S}_b^2$

Q.13 For sufficiently large values of N and M, the $V(\overline{y}_{nm})$ can take the following reduced form-

(a) $V(\overline{y}_{nm}) = \frac{S_b^2}{n} + \frac{\overline{S}_w^2}{nm}$ (b) $V(\overline{y}_{nm}) = \frac{S_b^2}{m} + \frac{\overline{S}_w^2}{nm}$

(c) $V(\overline{y}_{nm}) = \frac{S_b^2}{nm} + \frac{\overline{S}_w^2}{n}$ (d) $V(\overline{y}_{nm}) = \frac{S_b^2}{nm} + \frac{\overline{S}_w^2}{m}$

Q.14 While adopting two-stage sampling scheme, if n = N, then two stage sampling scheme reduces to

(a) Stratified random sampling (b) Cluster sampling
(c) Two stage cluster sampling (d) Systematic sampling

Q.15 Under certain conditions, the following two sampling schemes can be regarded as particular case of two stage sampling scheme.

(a) Cluster sampling and systematic sampling
(b) Cluster sampling and stratified random sampling
(c) Stratified random sampling and single random sampling
(d) None of the above

Q.16 The following expression is correct to estimate $V(\overline{y}_{nm})$ unbiasedly.

(a) $$\hat{V}(\overline{y}_{nm}) = \frac{(N-n)}{nN} s_b^2 + \frac{1}{N}\left(\frac{M-m}{mM}\right)\overline{s}_w^2$$

(b) $$\hat{V}(\overline{y}_{nm}) = \frac{N-n}{nN} s_b^2 + \frac{1}{n}\left(\frac{N-m}{mN}\right)\overline{s}_w^2$$

(c) $$\hat{V}(\overline{y}_{nm}) = \frac{N-n}{nM} \overline{S}_w^2 + \frac{1}{N}\left(\frac{M-m}{mM}\right)\overline{s}_b^2$$

(d) $$\hat{V}(\overline{y}_{nm}) = \frac{N-m}{nM} \overline{s}_w^2 + \frac{1}{n}\left(\frac{N-n}{nN}\right)\overline{s}_w^2$$

Q.17 In case of unequal S.S.U. and F.S.U., an unbiased estimator for population mean $\overline{Y}_N$ is defined as

(a) $\overline{y}_U = \frac{1}{n}\sum_i^n \left(\frac{\overline{M}_i}{\overline{M}}\right)\overline{y}_{imi}$ (b) $\overline{y}_U = \frac{1}{n}\sum_i^n \left(\frac{\overline{M}}{\overline{M}_i}\right)\overline{y}_{imi}$

(c) $\overline{y}_U = \frac{1}{n}\sum_i^n \left(\frac{M_i}{\overline{M}_n}\right)\overline{y}_{imi}$ (d) $\overline{y}_U = \frac{1}{n}\sum_i^n \left(\frac{nn_i}{\overline{M}}\right)\overline{y}_{imi}$

Q.18 Two phase sampling is also known as

(a) Two stage sampling (b) Double sampling
(c) Sub sampling (d) None

Q.19 In multiphase sampling scheme, the sampling units are

(a) Same in all the phases
(b) Defferent in all the phase
(c) Smaller in size in succeeding phases
(d) Larger in preceeding phases

Q.20 In multistage sampling scheme, the sampling units

(a) Remain same in size at different stages
(b) Remain smaller in size in succeeding stages
(c) Remain larger in size in succeeding stages
(d) Remain smaller in size in preceeding stages

7.6 Key Answers

7.6.1 True / False

1.	True	2.	False	3.	True
4.	True	5.	True	6.	True
7.	True	8.	False	9.	True
10.	True	11.	True	12.	False
13.	True	14.	True	15.	True
16.	True	17.	True	18.	False
19.	False	20.	True	21.	False
22.	False	23.	False	24.	True
25.	True				

7.6.2 Fill in the Blanks

1. Sampling variance
2. Enumerated
3. Population
5. Representative
6. Two stage sampling, Sub-sampling
7. First stage, Primary stage
8. Multistage

9. Sampling, Independently
10. Large sample
11. Random Stratified random
12. Single stage
13. Unbiased
14. S^2_b, $\overline{S}^2_w$
15. $(\overline{Y}_i - \overline{Y}_N)$
16. $(M-1)$
17. n, nm
18. Sub-Sampling
19. Stratified random sampling
20. Two stage or sub-sampling
21. Two stage or sub-sampling
22. s^2_b, $\overline{s}_w^{\ 2}$
23. $(Mi / \overline{M})$
24. Same
25. Different
26. Same
27. Different
28. Larger

7.6.3 Multiple Choice Questions and Answers

1.	b	2.	a	3.	d
4.	d	5.	b	6.	a
7.	a	8.	b	9.	a
10.	a	11.	b	12.	b
13.	a	14.	a	15.	b
16.	a	17.	a	18.	b
19.	a	20.	b		

8

Varying Probability Samplings

8.1 True / False

Q.1 A sampling procedure in which the units are selected with probability proportional to their some size measure, is known as sampling procedure with probability proportional to size (P.P.S.)

Notations: Let there be a population of N units say Y_i, (i = 1,2, ... N), Y_N with their selection probabilities Pi (i=1, 2........... , N). Here Pi $\neq$ Pj $\forall$ i $\neq$ J. = 1, 2, ... N. Let y_i, (i = 1, 2, ...n) be the sample drawn from the population with probability of selection pi (i = 1, 2 -, n).

Q.2 An unbiased estimator t_n for the population mean $\overline{Y}_N$ is defined as $t_n = \frac{1}{n}\sum_{i=1}^{n}\frac{yi}{(Npi)}$ under a sampling scheme "Probability proportional to size with replacement (PPSWR)".

Q.3 An expression to work out the sampling variance of an unbiased estimator t_n under PPSWR scheme is written as

$$V(t_n)_{PPSWR} = \frac{1}{nN^2}\left(\sum_{i}^{N}\frac{Y_i^2}{pi} - Y\right)^2$$ where Y is the population total.

Q.4 When $P_i \; \alpha y_i \; ie Pi = Yi / Y$ then $V(\hat{Y})_{PPSWR} = 0$. Thus, when $Pi \, \alpha \, yi$ then PPSWR is better then SRSWOR.

Q.5 To select a sample under PPSWR, the two methods of selection are as-

1. Cumulative total method
2. Lahiri method of sample selection

Q.6 The main draw back of cumulative method of sample selection is that the successive cumulative total of X's is time consuming and tedious.

Q.7 If population size N $\leq$ 5 ie (N is small) then cumulative method of sample selection in not preferable.

Q.8 $\hat{V}(\hat{Y})$ PPSWR can be worked out from the expression

$$\hat{V}(\hat{Y})_{PPSWR} = \frac{1}{n(n-1)}\left[\sum_{i=1}^{n}\left(\frac{Y_i}{p_i}\right)^2 - n\hat{Y}^2_{PPSWR}\right]$$

Q.9 Des Raj ordered estimators are biased estimators of population total Y.

Q.10 Des Raj ordered estimators (t_i, t_j, i $\neq$ J = 1, 2, _____ n) are pairwise correlated.

Q.11 If t_1, t_2, _____ t_n are all unbiased estimator of population total and are pairwise uncorrelated then the estimator

$\bar{t} = \frac{1}{n}\sum_{i=1}^{n} t_i$ is unbiaased for Y and is known as Des Raj estimator under PPSWOR.

Q.12 In case of PPSWR scheme, the relation among sampling variance of Des Raj ordered estimators Say (t_1, t_2, _____ , t_n) exists as $V(t_1) \leq V(t_2) \leq \ldots \leq V(t_n)$

Q.13 $V(\bar{t})_{POSWR} \geq V(\bar{Y}_{PPSWR})$

Q.14 The inclusion probability of a sampling unit in the sample has been defined by Harvitz Thomson as $\pi i := P(Yi)$: It is the ultimate probability of selecting the ith unit in the sample of size n.

Q.15 The ultimate joint probability of sampling units (Y_i, Y_j) to be included in the sample of size n is defined as $\pi ij = P(y_i, y_i)$

Q.16 In "sampling scheme" the probability of selection is associated with the individual sampling unit.

Q.17 In "Sampling design" the probability of selection is associated with the selection of sample not with individual units.

Q.18 For a given sampling scheme, there exists a sampling design and vice versa.

Q.19 Let's denotes the sample. Then inclusion probability of ith unit is defined as

$$\pi_i = \sum_{i \in s}^{n} P(s) \text{ and } \pi ii = \sum_{s \in i,j}^{n} P(s)$$

Where P(s) denotes the selection probability of the sample s.

Q.20 Harvitz Thomson estimator $\hat{Y}_{HT}$ for population total Y under PPSWOR scheme is defined as

$$\hat{Y}_{HT} = \sum_{i=1}^{n} y_i / \pi_i$$

Q.21 The estimator $\hat{Y}_{HT}$ gives biased estimate of Y.

Q.22 The expression for $V(\hat{Y}_{HT})$ is given as-

$$V(\hat{Y}_{HT}) = \sum_{i=1}^{N} \left(\frac{1-\pi_i}{\pi_i} \right) Y_i^2 + \sum_{i \pm j}^{N} \left(\frac{\pi_{ij} - \pi_i \pi_j}{\pi_i \pi_j} \right) y_i \, y_j$$

Q.23 Sometime the estimator $\hat{V}(\hat{Y}_{HT})$ takes negative value for some samples, which creates problems in interpreting the reliability of the estimates.

Q.24 The variance expression $\hat{V}(\hat{Y}_{HT})$ as proposed by Yates & Grundy (1953) is given as

$$\hat{V}_{YG}(\hat{Y}_{HT}) = \sum_{j>i}^{n} \left(\pi_i \pi_j - \pi_{ij} \right) \left(\frac{y_i}{\pi_j} - \frac{y_i}{\pi_j} \right)^2$$

Q.25 An expression to work out $\hat{V}_{YG}(\hat{Y}_{HT})$ is written as

$$\hat{V}_{YG}(\hat{Y}_{HT}) = \sum_{i=1}^{n} \sum_{j>i}^{n} \left(\pi_i \pi_j - \pi_{ij} \right) \left(\frac{y_i}{\pi_i} - \frac{y_j}{\pi_j} \right)^2$$

Q.26 In case of PPSWOR, if $y_i \alpha \pi_i$ then sampling variance of an estimator $\hat{Y}_{HT}$ reduces to zero and we get an efficient estimate.

Q.27 If $\pi i \, \alpha \, x i$, then sampling scheme is known as πPS sampling scheme.

Q.28 In PPSWOR scheme if $\pi_{ij} > 0 \; \forall \; i \neq j = 1, 2,$ _____N, then sampling variance $V(\hat{Y}_{HT})$ can be estimated unbiasedly.

Q.29 In Midjuno-Sen sampling scheme, the inclusion probability $\pi_{ij} > 0 \; \forall \; i \neq j = 1, 2$_____ N , and we get an unbiased estimate of sampling variance of an estimator under this scheme.

Q.30 In Mid-Zuno-Sen sampling scheme, $(\pi_i \pi_j - \pi_{ij})$ alwasys and hence is always positive in this sampling scheme.

Q.31 $\hat{V}_{YG}(\hat{Y}_{HT})$ always provides non-negative estimate of sampling variance of the estimateor $\hat{Y}_{HT}$.

Q.32 When probability of selection of different samples varies from sample to sample then the sampling procedure is known as "varying probability sampling scheme."

Q.33 In case of Des Raj sequence of etimators, an unbiased estimator t of Y^2 is defined as

$$t = \frac{1}{n(n-1)} \sum_{i \neq j}^{n} ti\ tj$$

8.2 Fill in the Blanks

Q.1 Under PPSWR sampling scheme, an unbiased estimator t_n for population mean $\overline{Y}_N$ is defined as

$$t_n = \frac{1}{n} \sum_{i=1}^{n} \frac{yi}{(.......)}$$

Q.2 $V(t_n)_{PPSWR}$ can be expressed as

$$V(t_n)_{PPSWWR} = t_n = \frac{1}{n} \sum_{i=1}^{n} \frac{1}{nN^2} (..................)^2$$

Q.3 When $Pi \alpha Yi$ then $V(\hat{Y})_{PPSWR} =$ ____________

Q.4 Cummulative total method is used to select a sample under __________ scheme.

Q.5 Lahiri method for sample selection is used under __________ scheme.

Q.6 When $N \leq 5$, then cummulative method of sample selection is ________

Q.7 Des Raj ordered estimators are _________ estimators of population total Y.

Q.8 Des Raj ordered estimators (t_i, t_j, $i \neq j = 1, 2,$ __________, n) are pairwise _________

Q.9 If (t_i, i =1, 2, ________,n) are Des Raj ordered estimators of population total Y, then the estimator

$t = \frac{1}{n} \sum_{i=1}^{n} t_i$ is an _________estimator for Y under PPSWOR scheme.

Q.10 In case of PPSWR scheme, the relation among ____________ of Des Raj ordered estimators exist as

$$V(t_1) \leq V(t_2) \leq .... \leq V(t_n)$$

Q.11 $\pi_i = P(Y_i)$ has been defined as ____________ of a sampling unit by Harvitz Thomson.

Q.12 $\pi_{ij} = P(Y_i\, Y_j)$ denote the joint ______________ probability of sampling units (Yi, Yj)

Q.13 The probability of selection of individual sampling units, is considered in "______________"

Q.14 The probability of selection of samples, is considered in "__________"

Q.15 For a given sampling scheme, there exists a "__________" and vice-versa.

Q.16 Let P(s) denotes the sampling design of the sample s then inclusion probability of ith unit in the sample s in defined as $\pi i =$ ____________

Q.17 Let π_{ij} denotes the inclusion probability of the sampling units (i, j) then π_{ij} is defined as $\pi_{ij} = \sum_{i,J \in S}$ ____________

Q.18 The estimator $\hat{Y}_{HT}$ for population total Y under PPSWOR scheme is difined as $\hat{Y}_{HT} = \sum_{i=1}^{n}..........$

Q.19 The estimator $\hat{Y}_{HT}$ gives an ____________ estimate of Y.

Q.20 The expression for $V(\hat{Y}_{HT})$ is written as

$$V(\hat{Y}_{HT}) = \sum_{i=1}^{N}\left(\frac{1-\pi_i}{\pi_i}\right)...... + \sum_{i \neq j}^{N}\left(\frac{\pi_{ij} - \pi_i\, \pi_j}{\pi_i\, \pi_j}\right)......$$

Q.21 The variance expression $V(\hat{Y}_{HT})$ as proposed by Yates and Grundy (1953) is written as

$$V_{YG}(\hat{Y}_{HT}) = \sum_{j>i}^{N}(\pi i\, \pi j - \pi ij)(.......)^2$$

Q.22 An expression to work out $\hat{V}_{YG}(\hat{Y}_{HT})$ is written as

$$\hat{V}_{YG}(\hat{Y}_{HT}) = \sum_{i=1}^{n}\sum_{j=i}^{n}\left(\frac{\pi_i\,\pi_j - \pi_{ij}}{\cdots\cdots\cdots\cdots}\right)\left(\frac{Y_i}{\pi_i} - \frac{Y_i}{\pi_j}\right)^2$$

Q.23 In case of PPSWOR scheme, $y_i\ \alpha\ \pi_i$ then $V(\hat{Y}_{HT})$ reduces to ______

Q.24 If $\pi_i\ \alpha\ x_i$, then sampling scheme is known as ____________ sampling scheme.

Q.25 In PPSWOR scheme if $\pi_{ij} > 0\ \forall\ i \neq j = 1, 2,$ ____________, N, then $V(\hat{Y}_{HT})$ can be estimated ________

Q.26 In Midjuno-Sen sampling scheme, $\pi_{ij} > 0\ \forall\ i \neq j = 1, 2, \ldots\ldots, N$. We can get an ____________ estimate of V (estimator).

Q.27 In Midjuno-Sen sampling scheme, $(\pi i\,\pi j - \pi ij) > 0$ always and hence $V(\hat{Y})$ is found always ____________ in this scheme.

Q.28 $\hat{V}_{YG}(\hat{Y}_{HT})$ always provides ____________ estimate of $V_{YG}(\hat{Y}_{HT})$.

Q.29 When sampling design (P(S)) varies from sample to sample then sampling scheme is known as "____________" sampling scheme.

Q.30 In case of Des Raj sequence of ordered statistics (t_1,i = 1,2, ______ n), an unbiased estimator t_n for Y^2 is defined.

$$t_n = \frac{1}{n(n-1)}\sum_{i \pm j}^{n} \cdots\cdots\cdots\cdots$$

8.3 Multiple Choice Questions and Answers

Q.1 Under PPSWR scheme, an unbiased estimator t_n for population mean $\overline{Y}_N$ is defined as.

(a) $t_n = \frac{1}{n}\sum_{i}^{n}\frac{Y_i}{p_i}$

(b) $t_n = \frac{1}{n}\sum_{i=1}^{n}\frac{y_i}{Np_i}$

(c) $t_n = \frac{1}{n}\sum_{i}^{n} y_i\, p_i$

(d) $t_n = \frac{1}{n}\sum_{i}^{n}\frac{y_i\, p_i}{N}$

Q.2 An expression for sampling variance of an unbiased estimator t_n is defined as.

(a) $V(t_n)_{PPSWR} = \frac{1}{nN^2}\left(\sum_{i}^{N}\frac{Y_i^2}{Pi} - Y\right)^2$

(b) $V(t_n)_{PPSWR} = \frac{1}{nN}\left(\sum_{i}^{N}\frac{y_i^2}{Pi} - Y\right)^2$

(c) $V(t_n)_{PPSWR} = \frac{1}{nN^2}\sum^{N}\left(\frac{y_i}{p_i} - Y\right)^2$

(d) $V(t_n)_{PPSWOR} = \frac{1}{nN}\sum_{i}^{N}(P_i Y_i - Y)^2$

Q.3 In PPSWR scheme if $Pi \alpha\, yi$ then $V(\hat{Y})$ is equal to

(a) $\frac{1}{N}$ (b) $\frac{N}{n}$

(c) Zero (d) ¥

Q.4 Cummulative method of sample selection is applicable in

(a) PPSWR scheme (b) PPSWOR scheme

(c) Midzuno-Sen scheme (d) None of these

Q.5 Lahiri method of sample selection in applicable in-

(a) PPSWOR scheme

(b) PPSWR scheme

(c) Midzuno-Sen scheme

(d) Sampling for proportion and percentage

Q.6 For $N \leq 5$, cummulative method of sample selection is

(a) Preferable (b) Not preferable

(c) Not possible (d) None

Q.7 $\hat{V}(\hat{Y})$ PPSWR can be worked out from the expression-

(a) $\frac{1}{n(n-1)}\left[\sum_{i}^{n}\left(\frac{y_i}{p_i}\right)^2 - n\hat{Y}^2\right]$

(b) $\frac{1}{n(n-1)}\left[\sum_{i}^{n}(p_i\, y_i)^2 - n\hat{Y}^2\right]$

(c) $\frac{1}{n(n-1)}\left[\sum_{i}^{n}\left(p_i\, y_i - \hat{Y}^2\right)\right]$

(d) $\frac{1}{n(n-1)}\left[\sum_{i=1}^{n}\left(\frac{y_i}{p_i} - \hat{Y}\right)^2\right]$

Q.8 Des Raj ordered estimators give-

(a) An unbiased estimate of $\overline{Y}_N$
(b) An unbiased estimate of Y
(c) An biased estimate of Y
(d) An biased estimate of $\overline{Y}_N$

Q.9 Des Raj ordered estimators (ti, tj), $i \neq j$, = 1, 2, __________, n) are

(a) Pairwise uncorrelated
(b) Pairwise correlated
(c) Pairwise independent
(a) Both (a) and (c)

Q.10 If (ti, tj, $i \neq j$) = 1, 2, __________ n are Des Raj order estimators of population total Y then the estimator $\overline{t}_n = \frac{1}{n}\sum_{i=1}^{n} t_i$ under PPSWOR scheme gives.

(a) An unbiased estimate of $\overline{Y}_N$
(b) An unbiased estimate of Y
(c) An biased estimate of Y
(d) An biased estimate of $\overline{Y}_N$

Q.11 In PPSWR for Des Raj ordered estimators, the following relation exists-

(a) $V(\overline{t}) = V(\hat{Y})$
(b) $V(\overline{t}) > V(\hat{Y})$
(c) $V(\overline{t}) \leq V(\hat{Y})$
(d) Can not be said with surity

Q.12 $\pi i = P(Y_i)$ denotes the Probability of selection of unit Y_i which has been given by ________

(a) Des Raj
(b) Harvitz Thomson
(c) P. C. Mahalanobias
(d) C. R. Rao

Q.13 $\pi_{ij} = P(y_i, y_j)$ joint denotes the probability of selecting the units (y_i, y_j, $i \neq j$, = 1,2,________ ,n) has been given by

(a) Prof. C. R. Rao
(b) Des Raj
(c) Harvitz Thomson
(d) Prof. R. A. Fisher

Q.14 The probablility of selecting an individual unit is related with the term

(a) Sampling scheme
(b) Sampling design
(c) Sampling strategy
(d) Sampline space

Q.15 The probability of selection of samples is related with the term

(a) Sampling strategy (b) Sampling scheme

(c) Sampling design (d) Sampling frame

Q.16 For a given sampling scheme, there exists a

(a) Sampling design (b) Purposive sampling

(c) Sample (d) Sampling variance

Q.17 For a given sampling design P(S), the inclusion probability πi for ith unit in the sample S, is defined as

(a) $\pi_i = \sum_{i \in s}^{n} P(s)$ (b) $\pi_i = \sum_{i \notin s}^{n} P(s)$

(c) $\sum_{i \in s}^{N} P(S)$ (d) None

Q.18 In clusion probability π_{ij} of the units (y_i, y_j, $i \neq j = 1, 2, \ldots\ldots\ldots, n$) jointly for a given sampling design P(s) is defined as

(a) $\pi_{ij} = \sum_{i \neq j \in S}^{n} P(s)$ (b) $\sum_{i > j \in S}^{n} P(s)$

(c) $\sum_{i \in S}^{N} P(s)$ (d) None

Q.19 Harvitz-Thomson estimator $\hat{Y}_{HT}$ for population total Y under PPSWOR scheme is defined as

(a) $\hat{Y}_{HT} = \sum_{i=1}^{n} \frac{y_i}{p_i}$ (b) $\hat{Y}_{HT} = \sum_{i=1}^{n} \frac{y_i}{\pi_i}$

(c) $\hat{Y}_{HT} = \sum_{i=1}^{n} p_i\, y_i$ (d) $\hat{Y}_{HT} = \sum_{i=1}^{n} \pi_i\, y_i$

Q.20 The estimator $\hat{Y}_{HT}$ provides the estimate of Y as

(a) Biased estimate (b) Unbiased estimate

(c) Sufficient estimate (d) Consistent estimate

Q.21 If $\pi_i \propto y_i$ in case of PPSWOR scheme then $V(\hat{Y}_{HT})$ reduces to-

(a) Zero (b) Negative quantity

(c) Positive quantity (d) ∞

Q.22 If $\pi_i \alpha x_i$, then the smapling scheme is known as

(a) P P S sampling scheme
(b) π P S sampling scheme
(c) Purposive sampling scheme
(d) Large sampling scheme

Q.23 In PPSWOR scheme, if $\pi ij > 0 \forall i \neq j = 1, 2, \_\_\_\_, N$ then $V(\hat{Y}_{HT})$ can be estimated

(a) Biasedly
(b) Consistently
(c) Unbiasedly
(d) Efficiently

Q.24 In Midzuno-Sen sampling scheme, $\pi ij > 0 \forall i \neq j = 1, 2, \_\_\_\_, N$, V (Estimator) can be estimated

(a) Biasedly
(b) Unbiasedly
(c) Consistently
(d) Efficiently

Q.25 In Mid-Zuno-Sen sampling scheme, the value of $(\pi i \pi j - \pi ij)$ is always

(a) Less then zero
(b) Greator then zero
(c) Equal to zero
(d) Equal to 0.5

Q.26 $\hat{V}_{YG}(\hat{Y}_{HT})$ always provides

(a) Non-negative estimate
(b) Negative estimate
(c) Estimate equal to zero
(d) Estimate equal to ∞

Q.27 When P (S) varies from sample to sample then sampling scheme is known as

(a) Varying probability sampling scheme
(b) Probability proportional to size sample scheme
(c) Purposive sampling scheme
(d) πP S sampling scheme

Q.28 In case of Des Raj sequence of estimators, an unbiased estimator "t_n" of Y^2 is defined as-

(a) $t_n = \frac{1}{n}\sum_{i \pm j}^{n} t_i t_j$
(b) $t_n = \frac{1}{n(n-1)}\sum_{i \pm j}^{n} t_i - t_j$
(c) $t_n = \frac{1}{(n-1)}\sum_{i \neq j}^{n} t_i t_j$
(d) None

8.4 Key Answers

8.4.1 True/False

1.	True	2.	True	3.	True
4.	True	5.	True	6.	True
7.	False	8.	True	9.	False
10.	Fasle	11.	True	12.	False
13.	False	14.	True	15.	True
16.	True	17.	True	18.	True
19.	True	20.	True	21.	False
22.	True	23.	True	24.	True
25.	True	26.	True	27.	True
28.	True	29.	True	30.	True
31.	True	32.	True	33.	True

8.4.2 Fill in the Blanks

1. Np_i
2. $\left(\sum_{i}^{N} \frac{y_i^2}{p_i} - Y\right)$
3. Zero
4. P P S W R
5. P P S W R
6. Preferable
7. Unbiased
8. Uncorrelated
9. Unbiased
10. Sampling Variance
11. Inclusion probability
12. Inclusion probability
13. Sampling scheme
14. Sampling design
15. Sampling design
16. $\sum_{i \in S}^{n} P(S)$

17. P(S)

18. $\frac{y_i}{p_i}$

19. Unbiased

20. Yi^2, Yi Yj

21. $\left(\frac{y_i}{\pi_i}-\frac{y_i}{\pi_j}\right)$

22. $\pi_i \pi_j$

23. Zero

24. πP S

25. Unbiasedly

26. Unbiased

27. Positive

28. Non-negative

29. Varying Probability

30. t_i t_j

8.3.3 Multiple Choice Questions and Answers

1.	b	2.	a	3.	c
4.	a	5.	b	6.	a
7.	a	8.	b	9.	d
10.	b	11.	c	12.	b
13.	c	14.	a	15.	c
16.	a	17.	a	18.	a
19.	b	20.	b	21.	a
22.	b	23.	c	24.	b
25.	b	26.	a	27.	a
28.	b				

9

Preliminaries and Basic Principles of Design of Experiments

9.1 True/ False

Q.1 An experiment is a means of getting an answer to the question which arises in the mind of an experimenter (reseacher).

Q.2 An experiment may be planned for not to compare the yield of several varieties of wheat.

Q.3 Design of experiment means to decide how the observations or measurements should be taken on experimental units to aswer a particular question in a valid, efficient and economic way.

Q.4 Design of an experiment and final analysis go together in the sense that if an experiment is properly designed there exists an appropriate way to analyse the data.

Q.5 Absolute experiment and comperative experiment are not the two categories in which the design of an experiment can be classified.

Q.6 An experiment through which an absolute value of some characteristic under study is determined, is known as an absolute experiment.

Q.7 Finding an average inteligence quotient (I.Q.) of a group of students is not an example of an absolute experiment.

Q.8 An experiment through which the effects of two or more than two treatments (fertilizers) under study on some population characterstics are compared, is known as a comparative experiment

Q.9 The smallest parts of the experimental material to which the treatment under consideration are applied and observations are taken, are not known as experimental units.

Q.10 In field experiments, the small plots of land and in hospital, the individual patients may be treated as an experimental units.

Q.11 Finite collection of all experimental units is known as experimental material.

Q.12 Variation in measurements taken on two adjacent experimental units treated in an identical way, is known as experimental error.

Q.13 An experimental error is always of random nature.

Q.14 An experimental error does not arise due to inherent variability in the experimental material.

Q.15 Experimental error may occure due to lack of uniformity in mothodology of conducting the experiment.

Q.16 Various objects (or procedures) to be compared in a comperative experiment, are termed as treatments.

Q.17 In field experimentation, the different fertilizers or different varieties of a crop may be named as treatments.

Q.18 An experimental error provides the basis for confidence to be placed in inference drawn from the experimental results.

Q.19 Estimation of an experimental error is never required to be used in defining F statistic.

Q.20 The control of an experimental error does not help in increasing the efficiency of experiment.

Q.21 An experimental error is controlled by the principle of "Local Control"

Q.22 An experimental errors can not be estimated by using the principle of "Replication"

Q.23 Randomization does not eliminate the human bias in an experiment.

Q.24 The precision of an experiment can be measured by an expression-

$$\text{Precision (Experiment)} = \frac{1}{\sigma^2_{\bar{x}n}} = \frac{n}{\sigma^2}$$

where n is the number of repetion of an observation taken on experimental unit.

Q.25 The precision of an experiment cannot be increased by increasing the number of replication

Q.26 The precision of an experiment cannot be increased by using the principle of local control.

Q.27 To get a valid conclusion from the experiment, it must be conducted in identical condition w. r. to all other variables except that understudy variable.

Q.28 Randomization eliminates the biasness of an experimenter done in an experiment.

Q.29 A linear function $C=\sum_{i=1}^{n} \ell i\, yi$ where l_i (i = 1, 2, ..., n) are given constants such that $\sum_{i=1}^{n} \ell_\ell = 0$, is known as contrast.

Q.30 The restriction $\sum_{i=1}^{n} \ell_i = 0$, does not make the contrast C to be able to make comparision among Y values.

Q.31 An expected value of C i. e E (C) $=M\sum_{i=1}^{n} \ell i$, is always equal to zero.

Q.32 The variance of contrast C is defined as $=V(C)=\sigma^2 \sum_{i}^{n} \ell i^2$.

Q.33 The condition for two contrasts $A=\sum_{i=1}^{n} \lambda_i x_i$ and $B=\sum_{i=1}^{n} M_i x_i$ to be orthogonal is that $\sum_{i=1}^{n} \lambda i\, Mi = 0$

Q.34 A contrast C defined as $C=\sum_{i}^{n} \lambda_i x_i, \sum_{i=1}^{n} \lambda_i = 0$ is said to be an elementary contrast if l_i have only two non zero element as ± 1.

Q.35 Let C_1 and C_2 be two contrasts defined as

$C_1 = y_1 + y_2 - y_3 - y_4$

$C_2 = y_1 - y_2 - y_3 + y_4$

Then C_1 and C_2 are both elementary contrasts and are orthogonal to each other.

Q.36 A graphic picture which shows the nature of soil fertility variation obtained in a uniformity trial by joining the points, of equal fertility through lines, is known as "fertility contour map".

Q.37 Uniformity trial does not give us some idea about the shaps and size of the experimental plots.

Q.38 To get the maximum precision in an experiment, the plots must be rectangular in nature.

Q.39 To get the maximum precision in an experiment, the long side of rectangular plots must be parallel to the direction of fertility gradient.

Q.40 While making blocks in an experiment, the blocks should be arranged one after other along the fertility gradient.

Q.41 In an experiment while applying the principle of local control, the blocks should not be made perpendicular to the direction of fertility gradient.

Q.42 The accuracy of a measurement signifies the closeness with which the measurement approchas to the true value.

Q.43 If the deviation of measurement from its true value is lesser then greater is the accuracy of measurement.

Q.44 The precision of the measurements, denotes the closeness with which a measurement approaches to the average value of these measurements.

Q.45 The lesser is the variability of the measurements from their average value, the greater is the precision of these measurements and their estimates.

Q.46 Professor R. A. Fisher is the pioneer statistician for the study of experimental designs.

Q.47 Replication does not mean the repetition of the treatments under investigation.

Q.48 Replication makes it possible to get more reliable estimate than that which is obtained through a single observation.

Q.49 The precision of an experiment in not directly proportional to the number of replication of treatments under investigation.

Q.50 The most important role of replication is to provide an estimate of the experimental error.

Q.51 The estimate of the experimental error is obtainde by considering the differences of two observations taken on plots receiving the same treatment in different replications.

Q.52 A general rule is not to use as many replications which provide at least 12 d.f. for error.

Q.53 There is a fact that the values of F statistic do not decrease rapidly beyond 12 error d. f.

Q.54 Usually, one should not use less than 4 replications in an experiment.

Q.55 An approximate minimum number of replication required to detect a specified difference a between two treatment means at a% level of significance is given by

$$r = \frac{2t_{\alpha}^{2}\sigma^{2}}{d^{2}}$$

Q.56 The process of assigning the different treatments to the various experimental units in a purely random way is not known randomisation.

Q.57 Randomisation is essential for a valid estimate of experimental error.

Q.58 Randomisation does not play a role to minimize the human bias occured in an experiment.

Q.59 The assumption of independence of errors in ANOVA model is fulfilled by the application randomisation technique in the experiment.

Q.60 The random nature of trhe error term involved in ANOVA model is assured by the application of principle of randomisation.

Q.61 The reduction of experimental error by dividing the relatively heterogeneous experimental area into homogeneous blocks, is known as local control.

Q.62 The efficiency of designing of an experiment can not be increased by applying the principle of local control.

Q.63 By applying the replication and randomisation jointly, an experimental error becomes estimable in a valid way.

Q.64 An experimental error can not be reduced by applying the confounding technique in a factorial experiment.

Q.65 An experimental error can be reduced by the use of one or more auxiliary variables in ANCOVA.

Q.66 By the proper choice of the size and shape of the experimental plots and blocks, an experimental error can be reduced.

9.2 Fill in the Blanks

Q.1 An experiment is a means to get an answer to a question which arises in the mind of __________ or researcher.

Q.2 An experiment may be planned to compare the yied potential of different __________ of wheat crop.

Q.3 Design of experiment means to decide how to take observations or measurements on __________ units.

Q.4 Design of an experiment and final __________ go together to get results for interpretation.

Q.5 Absolute experiment and __________ experiment are two categories in which the design of __________ can be classified.

Q.6 An experiment through which an absolute value of some characteristic under study is determined, is known as an __________ experiment.

Q.7 Finding an average intelligence quotent (I. Q.) of a group of students in an example of an __________ .

Q.8 An experiment through which the effects of two or more than two treatments (fertilizers) on certain characteristic of a crop, are compared is known as __________

Q.9 The smallest parts of an experimental material on which the treatments are applied and observation are taken, are known as __________

Q.10 In field experiment, the small plots of land and in hospital, the individual patients may be treated as an __________ units.

Q.11 A finite collection of experimental units is known as __________ material.

Q.12 The variations observed in measurements taken on two adjacent experimental units treated in an identical way is known as__________

Q.13 An experimental error is always of __________ nature.

Q.14 An experimental error may arise due to inherent variability in __________ units.

Q.15 Experimental error may occure due to lack of __________ in methodology of conducting the experiment.

Q.16 Various objects or procedures or varieties of a wheat crop to be compared through an experiment are known as __________

Q.17 In field experimentation, the different fertilizes applied or different varieties of a crop sown, may be named as __________

Q.18 An experimental error provides the basis for confidence to be placed in inference drawn from __________ results.

Q.19 Estimation of an experimental error is required to be used in defining the __________ statistic.

Q.20 The control of an experimental error helps in __________ the efficiency of an experiment.

Q.21 An experimental error is reduced by using the principle of __________

Q.22 An experimental error can be estimated by using the principle of __________

Q.23 Randomization eliminates the __________ in an experiment.

Q.24 The precision of an experiment can be measured by an expression written as.

Precision (Experiment) = __________

Q.25 The precision of an experiment can be increased by __________ the number of replication.

Q.26 The precision of an experiment can also be increased by using the principle of __________

Q.27 A valid conclusion based on observations taken from an experiment, can be obtained only when all the three __________ of design of experiment are followed in the experiment.

Q.28 A function $C = \sum_{i=1}^{n} \lambda_i y_i$ such that $\left(\sum_{i}^{n} \lambda_i\right) = 0$ as known as __________

Q.29 The condition $\sum_{i}^{n} \lambda i = 0$, means the contrast C to be able to make __________ among y values.

Q.30 An expected value of C i.e E (C) = __________

Q.31 The condition for two contrasts $A = \sum_{i}^{n} \lambda_i y_i$ and $B = \sum_{i}^{n} \mu_i y_i$ to be orthogonal to each other is that $\sum_{i=1}^{n} \cdots\cdots = \cdots\cdots$

Q.32 The variance of contrast C is given as $V(C) = \sum_{i}^{n} \cdots\cdots\cdots$

Q.33 The condition for two contrasts $A = \sum_{i=1}^{n} \lambda_i X_i$ and $B = \sum_{i=1}^{n} \mu_i X_i$ to be non orthogonal is that $\sum_{i=1}^{n} \lambda_i \mu_i$ __________

Q.34 A contrast C defined as $C = \sum_{i=1}^{n} \lambda_i x_i$, $\sum_{i=1}^{n} \lambda_i = 0$ is said to be an __________ contrast if λi's have only two non zero dements as __________

Q.35 Let C_1 & C_2 be two contrasts defined as

$C_1 = Y_1 + Y_2 - Y_3 - Y_4$ and $C_2 = Y_1 - Y_2 - Y_3 + Y_4$

Then (i) C1 and C2 are both __________ contrasts

(ii) Both are __________ to each other.

Q.36 A graphic picture showing the nature of soil fertility variation obtained through uniformity trial is known as __________

Q.37 Uniformity trial gives us an idea about the shape and size of the __________

Q.38 To get the maximum precision in an experiment, the plots must be __________ in nature.

Q.39 To get the maximum precision in an experiment, the long side of the rectangular plot must be __________ to the direction of fertility gradient.

Q.40 While making blocks, in an experiment, the blocks should be arranged one after other __________to the fertility gradient.

Q.41 While applying the principle of local control, the blocks should made __________ to the direction of fertility gradient.

Q.42 The accuracy of a measurement signifies the closeness with which the measurement aproaches to its __________

Q.43 If the deviation of measurement from its true value is lesser then __________ is the of measurement is larger.

Q.44 The precision of the measurements denotes the closeness with which these measurements approach to their __________ value

Q.45 The lesser is the deviation of measurements from their average value, the greater is the .... of these measurements.

Q.46 The pioneer statistician who developed the theory of design of experiment is Prof __________

Q.47 Replication means the __________ of the treatments under investigation in an experiment.

Q.48 Replication means it possible to get a more reliable of parametor than that which is obtained through a single observation.

Q.49 The precision of an experiment is directly proportional to the number of __________ of treatments under investigation

Q.50 The most important rule of replication is to provide an __________ of the __________ error.

Q.51 The difference (deviation) of two observations taken on plots receiving the same treatment in different replications, is known as __________

Q.52 A general rule is to use as many replications which provides at least __________ d. f. for error.

Q.53 There is fact that the values of F statistic do not decrease repidly beyond __________ d.f. for error.

Q.54 Usually, one should not use less then __________ replications in an experiment.

Q.55 Minimum number of replications in an experiment required to detect a specified difference d between two treatment means at $\alpha\%$ level of significance is given by-

r =..............

Q.56 The process of assining the different treatments to the various experimental units purely in a random way is known as __________

Q.57 Randomisation is essential for valid estimation of __________

Q.58 Randomisation play an important role in minimising the __________ occured in an experiment.

Q.59 Random nature of an error term involved in an ANOVA model is maintained by the application of principle known as __________

Q.60 An assumption of independence of experimental error in an ANOVA model in fulfilled by the application of __________ technique in the experiment.

Q.61 The process of dividing the heterogeneous experimental area (material) in to homogeneous blocks is known as the principle of __________

Q.62 The principle of local control followed in an experiment __________ the efficiency of the experiment.

Q.63 By the joint application of replication and randomisation an __________ can be estimated in a __________ way.

Q.64 Confounding technique in factorial experiment is applied to reduce the __________ error.

Q.65 An experimental error can be reduced by using one or more __________ in ANCOVA.

Q.66 By reducing the block size in experiment, an __________ error can be reduced.

Q.67 Another name of critical difference in design of experiment is also known as __________

Q.68 Greater homogeneity within the blocks in an experiment is maintained using the principle of __________

Q.69 Greater homogeniety within the blocks can be maintained by reducing the size of __________

Q.70 The two designs D_1 and D_2 are said to be equally efficient if the efficiency of D_1 w.r. to D_2 is equal to __________

Q.71 Each contrast among K treatments has __________ d.f.

Q.72 The maximum possible number of orthogonal contrasts among K treatments is equal to __________

9.3 Multiple Choice Questions and Answers

Q.1 An experiment is a means to get an answer to a question which arises in the mind of __________

(a) Experimenter (b) Researcher
(c) Scientist (d) All the above

Q.2 An experiment may be planned to compare the yield potential of different varities of

(a) Wheat crop (b) Paddy crop
(c) Sugarcane crop (d) All the above

Q.3 Design of experiment means to decide how to take observation or measurement on

(a) Experimental units (b) Sampling units
(c) Plants (d) Plots

Q.4 Design of experiment and final analysis go together to get of results for

(a) Interpretation (b) Recommendation
(c) Writing (d) None of the above

Q.5 Absolute experiment and comperative experiment are two categories in which the design of experiment can be

(a) Divided (b) Classified
(c) Categorised (d) Both (b) and (C)

Q.6 An experiment through which an estimated value of some characteristic under study obtained is known as

(a) Comparative experiment (b) Absolute experiment
(c) Factorial experiment (d) None

Q.7 Finding an average of inteligence quotent (I. Q) of a group of students is an example of

(a) Absolute experiment (b) Simple experiment
(c) Factorial experiment (d) Both (a) & (b)

Q.8 An experiment in which the effects of two or more than two fertilizers on a certain crop are compared, is known as

(a) Factorial experiment (b) Simple experiment
(c) Comperative experiment (d) Both (a) and (c)

Q.9 The smallest parts of an experimental area on which treatments are applied, is known as

(a) Experiment units (b) Sampling units
(c) Experimental material (d) None

Q.10 Deviation observed in measurements taken on two adjacent experimental units treated alike is known as

(a) Sampling error (b) Experimental error
(c) Measuremental error (d) Error due to investigator

Q.11 An experimental error is always of the nature like

(a) Random (b) Erratic
(c) Systematic (d) Both (a) and (b)

Q.12 Various objects or procedures or varieties of a crop, to be compared through an experiment are generally known as

(a) Experimental units (b) Treatments
(c) Replications (d) None of the above

Q.13 In field experiment, different fertilizers or varieties of a crop used in general can be named as

(a) Treatments (b) Experimental units
(c) Replication (d) All the above

Q.14 Estimation of an experimental error is required to define the statistic known as

(a) F statistic (b) X^2 statistic
(c) T statistic (d) Z statistic

Q.15 An experimental error can be reduced by using

(a) Principle of local control
(b) Using one or more auxiliary variables
(c) Confounding technique
(d) All the three techniques

Q.16 An experimental error can be estimated using the principle of

(a) Randomisation (b) Replication
(c) Local control (d) Least square

Q.17 Randomisation technique eliminates in design of experiment the

(a) Experimental error (b) Human bias
(c) Non sampling error (d) Both (a) and (c)

Q.18 The precision of an experiment with r replications can be worked out by using the expression written as

(a) $\frac{r}{\sigma 2}$ (b) $\frac{\sqrt{r}}{\sigma}$

(c) $\frac{2r}{\sigma 2}$ (d) $\frac{\sqrt{2r}}{\sigma}$

Q.19 Precisions of an experiments can be increased by

(a) Increasing the number of replication
(b) By using the local control technique
(c) By reducing the block size through confounding
(d) Using all the above three

Q.20 A valid conclusion for a result obtained from an experimental design can be obtained only when

(a) Principle of randomisation is followed.
(b) Principle of replication is followed.

(c) Principle of local control is followed.
(d) All the three principles are followed.

Q.21 A function $C=\sum_{i=1}^{n}\lambda_i y_i$ such that $C=\sum_{i=1}^{n}\lambda_i=0$ is known as

(a) Elementary contrast (b) Contrast
(c) Orthogonal contrast (d) None

Q.22 The condition for two contrasts, $A=\sum_{i=1}^{n}\lambda_i y_i$ and $B=\sum_{i=1}^{n}\mu_i y_i=0$ to be orthogonal to each other is

(a) $\sum_{i}^{n}\lambda_i \mu_i=0$ (b) $\sum_{i=1}^{n}\lambda_i \mu_i=1$

(c) $\left(\sum_{i=1}^{n}\lambda_i \mu_i\right)^2=1$ (d) None

Q.23 The variance of contrast $C=\sum_{i}^{n}\lambda_i y_i$ is given by

(a) $V(C)=\sigma^2\sum_{i=1}^{n}\lambda_i^2$ (b) $V(C)=\sigma^2\sum_{i=1}^{n}\lambda_i$

(c) $V(C)=\sigma^2\left(\sum_{i}^{n}\lambda_i\right)^2$ (d) $V(C_i)=\sum_{i}^{n}\sigma_i^2\lambda_i^2$

Q.24 A contrast C defined as $C=\sum_{i}^{n}\lambda_i y_i$ is said to be an elementary contrast when

(a) $\sum_{i=1}^{n}\lambda i=0$

(b) λi have only two non zero values as ± 1

(c) λi have only two values 0 & 1

(d) Both (a) and (b)

Q.25 Let C_1 and C_2 be two contrasts defined as
$C_1 = Y_1+Y_2 - Y_3-Y_4$, $C_2 = Y_1-Y_2-Y_3+Y_4$ then C_1 and C_2 both are

(a) Elementary contrasts
(b) Non orthogonal to each other
(c) Orthogonal to each other
(d) Both (a) and (c)

Q.26 The uniformity trial give us an idea about

(a) Shape and size of an experimental plot
(b) Picture of variation of soil fertility
(c) Direction of fertility gradient of soil
(d) Both (a) and (b)

Q.27 To get the maximum precision in an experiment

(a) The experimental plots must be in rectangular shape
(b) The experimental plots must be in square shape
(c) The longer side of plots must be parallel to fertility gradient
(d) Both (a) and (c)

Q.28 While making blocks in an experimental design, the blocks

(a) Must be arranged one after other along the fertility gradient
(b) Must be arranged one after other perpendicular to fertility gradient
(c) Must be made such that longer sides must be perpendicular to fertility gradient
(d) Both (a) and (c)

Q.29 While applying the principle of local control, the blocks

(a) Must be arranged one after other perpendicular to fertility gradient
(b) Must be prepared such that longer sides must be perpendicular to fertility gradient
(c) Must be arranged one after other along the fertility gradient
(d) Both (b) and (c)

Q.30 The accuracy of measurement signifies the closeness with which they approach to their

(a) Average value (b) Median value
(c) True value (d) Both (b) and (c)

Q.31 The precision of the measurements denote the closeness with which they approach to their

(a) True values (b) Average value
(c) Median value (d) Model value

Q.32 The pioneer statistician who originated the theory of design of experiment is

(a) Prof. P.C Mahalanoby (b) Prof. R.A. Fisher
(c) Prof. J. Neyman (d) Prof. C.R. Rao

Q.33 Replication makes it possible to

(a) Get more reliable estimate of experimental error
(b) Get an estimate of error variance
(c) Increase the precision of an experiment.
(d) Get all the above three

Q.34 The deviation of two observations taken on plots receiving the same treatment in different replications is known as

(a) Error variance
(b) Experimental error
(c) Mean deviation of observation
(d) Standard error of treatment mean

Q.35 A general rule is _______ to use as many replications which provides

(a) Atleast 12 d. f of error (b) Minimum 12 error d. f.
(c) Error df ≥ 12 (d) All the above three

Q.36 Minimum number of replications in an experiment required to detect a specified difference between two treatment means at $\alpha\%$ level of significance is given by

(a) $r = \frac{t_\alpha^2 \sigma^2}{d^2}$ (b) $\frac{2 t_\alpha^2 \sigma^2}{d^2}$
(c) $\frac{t_\alpha^2 \sigma^2}{2 d^2}$ (d) $\frac{t_\alpha^2 \sigma^2}{d^2}$

Q.37 Randomisation plays an important role in minimisation of

(a) Human bias (b) Experimental error
(c) Error variance (d) All the above

Q.38 Assumption of independence of experimental error in an ANOVA model is fulfilled by the application of

(a) Randomisation (b) Replication
(c) Local control (d) All the above

Q.39 By the joint application of randomisation and replication

(a) An experimental error can be measured
(b) Error variance can be estimated in a valid way
(c) Precision of an experiment can be increased
(d) All the above three can be achieved

Q.40 Each contrast among K treatments has

(a) 1. d.f (b) (K-1) d.f
(c) (K-2) d.f (d) K. d.f

Q.41 The maximum possible number of orthogonal contrasts among K treatments is equal to

(a) K (b) K-1
(c) K-2 (d) None of the above

9.4 Key Answers

9.4.1 True/False

1.	True	2.	False	3.	True
4.	True	5.	False	6.	True
7.	False	8.	True	9.	False
10.	True	11.	True	12.	True
13.	True	14.	False	15.	True
16.	True	17.	True	18.	True
19.	False	20.	False	21.	True
22.	False	23.	True	24.	True
25.	False	26.	False	27.	True
28.	True	29.	True	30.	False
31.	True	32.	True	33.	True
34.	True	35.	True	36.	True
37.	True	38.	True	39.	True
40.	True	41.	False	42.	True
43.	True	44.	True	45.	True
46.	True	47.	True	48.	True
49.	False	50.	True	51.	True
52.	False	53.	True	54.	True
55.	True	56.	False	57.	True

58.	False	59.	True	60.	True
61.	True	62.	False	63.	True
64.	False	65.	True	66.	True

9.4.2 Fill in the Blanks

1. Investigator or Scientist
2. Varieties
3. Experimental
4. Analysis
5. Comparative
6. Absolute
7. Absolute experiment
8. Comparative experiment
9. Experimental units
10. Experimental units
11. Experimental
12. Experimental error
13. Random or erratic
14. Experimental
15. Uniformity
16. Treatments
17. Treatments
18. Experimental
19. F
20. Increasing
21. Local control
22. Replication
23. Human error
24. $\frac{r}{\sigma^2}$
25. Increasing
26. Local control
27. Principles

28. Contrast
29. Comparision
30. Zero
31. $\sum_{i=1}^{n} \lambda_i \mu_i = 0$
32. $\sigma 2 \sum_{i}^{n} \lambda_i^2$
33. $\sum_{i}^{n} \lambda_i \mu_i \neq 0$
34. Elementary
35. Elementary, orthogonal
36. Contour map
37. Experimental plots
38. Rectangular
39. Parallel
40. Parallel
41. Perpendicular
42. True value
43. Accuracy
44. Average
45. Precision
46. Prof. R. A. Fisher
47. Replication
48. Estimate
49. Replication
50. Estimate, Experimental
51. Experimental error
52. 12
53. 12
54. 4
55. $\frac{2t_{\alpha}^{2}\sigma^{2}}{d^{2}}$

56. Randomisation
57. Experimental error
58. Human bias
59. Randomisation
60. Randomisation
61. Local control
62. Increases
63. Experimental error, valid
64. Experimental error
65. Auxiliary variaties
66. Experimental error
67. Least significant difference
68. Local control
69. Blocks
70. One
71. One
72. (K-1)

8.3.3 Multiple Choice Questions and Answers

1.	d	2.	d	3.	a
4.	a	5.	d	6.	b
7.	d	8.	d	9.	a
10.	b	11.	d	12.	b
13.	a	14.	a	15.	d
16.	b	17.	b	18.	b
19.	d	20.	d	21.	b
22.	a	23.	a	24.	d
25.	d	26.	d	27.	d
28.	d	29.	d	30.	c
31.	b	32.	b	33.	d
34.	b	35.	d	36.	b
37.	a	38.	a	39.	d
40.	a	41.	b		

10

ANOVA (One Way) Classification

10.1 True/False

Q.1 According to prof. R. A. Fisher "ANOVA" is not the separation of variance assignable to one group of causes from the variance assignable to another group of causes.

Q.2 The basic purpose of ANOVA technique is to test the homogeneity (equality) of several means.

Q.3 The basic purpose of ANOVA technique is to test the equality of several estimates of population variances.

Q.4 Prof. R.A. Fisher introduced first time the term ANOVA in agricultural data.

Q.5 ANOVA technique is now frequently used in testing the linearity of fitted regression line.

Q.6 ANOVA technique is also used in testing the significance of correlation ratio η.

Q.7 One of assumptions of ANOVA technique is that the variate value y_{ij} should be of random nature otherwise their means μ_{ij} can not be estimated

Q.8 In ANOVA model, the variate value y_{ij} should follow additive law for the several effects of treatments and error component.

Q.9 The error term e_{ij} in ANOVA model should be randomly, independently and normally distributed with mean zero and constant variance $V(eij) = \sigma_e^2$

Q.10 We get an unbiased estimate of constant variance σ_e^2 even when the hypothesis of homogeneity of means is significant (rejected).

Q.11 The variates y_{ij} should be homosedastic in ANOVA model otherwise we cannot get independent estimate of common variance σ_e^2

Q.12 The variates y_{ij} should be normally distributed because it is necessary to apply the F test even if the variates are random.

Q.13 The d. f. due to various sources of variation in ANOVA model should not be of additive nature.

Q.14 If the null hypothesis is true then theoretically the value of F statistic is equal to unity otherwise it will be greater than unity.

Q.15 In case of defining F statistic by Cochran theorem, $\frac{st^2}{\sigma_e^2}$ and $\frac{s^2e}{\sigma_e^2}$ are independently distributed as X^2 with (t - 1) and (n - t) d.f respectively

Q.16 The ratio $F = \frac{s^2_t/(t-1)}{s_e^2/(n-t)} = F_{(t-1),(n-t)}$

Q.17 The treatments and error terms must follow normal distribution because only in that case (s_t^2/σ^2e) and $\left(\frac{s^2e}{\sigma^2e}\right)$ will be independently distributed as χ^2 distribution.

Q.18 F statistic is defined as the ratio of two dependent mean Chi-square statistic.

Q.19 The sum of squares S_T^2 and S_E^2 are add up to get total S.S. but mean square (s_t^2 and s_e^2) are not add to get total mean square.

Q.20 The technique though is known as ANOVA but in fact it is the technique of analysis of S. S.

Q.21 The statistical model is actually a linear function of effects of a number of factors (treatments) involved in an experiment along with one or more error terms.

Q.22 A model in which each of the factors has fixed effect and only error effect is random, is known as fixed effect model.

Q.23 A model in which some factors have fixed effect and some other factors have random effect is known as random effect model.

Q.24 A model in which all factors have random effect is known as random effect model.

Q.25 The main objective of the fixed effect model is to estimate the effect of different factors involved in an experiment.

Q.26 In a random effect model, the main objective is to estimate the variability among the effects of different factors.

Q.27 When a set of observations are distributed over the different levels of only factor then such type of distribution (Classification)of observations is known two way classification.

Q.28 Complete randomised design (C. R. D) is the example of one way classification.

Q.29 When a set of observations are classified w.r. a to two factors say A and B simultaneously at a number of levels t & r respectively then such type of classification is known as two way classification.

Q.30 The data put in the form of randomised complete block design (R. C. B. D) is an example of two way classified data.

Q.31 The model for one way classified data is written as

$$y_{ij} = \mu + \beta_i + e_{ij} \qquad i = 1, 2, \ldots\ldots t, \quad j = 1, 2, \ldots\ldots, ni$$

Q.32 In question 31, μ and β_i are estimated as

$$\hat{\mu} = \frac{1}{n}\sum_{i=1}^{t}\sum_{j}^{ni} y_{ij} = \bar{y}.., \quad \hat{\beta}_i = (\bar{y}_i - \bar{y}..)$$

Q.33 Let s^2_A denotes the mean square due to factor A, then $E(s^2_A) = \sigma^2 e + \phi(A_1, A_2, \ldots\ldots, A_K)$ where $\phi(A_i \quad i = 1, 2, \ldots\ldots, K)$ is a variance like function of A_{is}

Q.34 In question 33, the value of $\phi(A_i \quad i = 1, 2, \ldots\ldots, K)$ is equal to zero under Ho otherwise it is a positive quantity.

Q.35 When Ho is true the treatment mean square gives an unbiased estimate of error variance σ_e^2.

Q.36 When Ho is rejected in ANOVA test then in place of F statistic, the students t statistic is used to test the pairwise equality of treatment means.

Q.37 The critical difference between two treatment means $(\bar{t}_i, \bar{t}_i)$ is defined in case of C.R. D. with equal replication r as.

C. D. $(\bar{t}_i, \bar{t}_{i'}) = t_{\alpha/2}\sqrt{\frac{2}{r}sc^2}$ where $t_{\alpha/2}$ is the upper t value at $\alpha/2$ level of significance at error d. f.

10.2 Fill in the Blanks

Q.1 ANOVA is the technique of separation of __________ assignable to one group of causes from the __________ assignable to another group of causes

Q.2 The basic purpose of ANOVA is to test the __________ of several means.

Q.3 Prof __________ introduced first time the term ANOVA in agricultural data.

Q.4 ANOVA technique is also used in testing the linearity of __________ line.

Q.5 ANOVA technique can also be used in testing the significance of __________ ratio η.

Q.6 One of the assumptions of ANOVA technique is that the variate value y_{ij} should be of __________ nature.

Q.7 In ANOVA model, the variate value y_{ij} should follow the __________ law for several effects of treatments and error components.

Q.8 The error term e_{is} in ANOVA model should be randomly, __________ and __________ distributed

Q.9 The error term eij is normally distributed with mean __________ and constant __________ $\sigma^2 e$

Q.10 We get an __________ estimate of constant variance $\sigma^2 e$ even when Ho of homogeneity of means is significant.

Q.11 The variate value y_{ij} should be homosedastic in __________ model otherwise we can not get independent estimate of __________ variance $\sigma^2 e$

Q.12 The variates y_{ij} should be __________ distributed because it is necessary to apply the __________ statistic to test the homogeneity of several means.

Q.13 The d. f. due to various sources of variation in ANOVA model should follow the __________ law.

Q.14 If the null hypothesis Ho is true then theoretically the value of F statistic is equal to __________

Q.15 If s_t^2 s_e^2 and $\sigma^2 e$ are treatment M.S, error M.S and error variance respectively then the ratios $\frac{s^2 t}{\sigma^2 e}$ and $\frac{S^2 e}{\sigma^2 e}$ are __________ distributed as __________ distribution with (t- 1) and (n-1) d.f.

Q.16 The treatment effect $\propto_t$ and error effect e_{ij} must be __________ distribution.

Q.17 The F statistic is defined as the ratio of two __________ mean chisquare statistics.

Q.18 In one way classified data S S, (s_t^2 and s_e^2) are add up to get total SS but M.S. ($s^2 t$ & $s^2 e$) are __________ up to get total M.S.

Q.19 Although the technique is known as ANOVA but in fact it is the technique of analysis of __________

Q.20 The statistical model is actually a linear function of effects of a number of treatment (factors) along with one or more __________ terms.

Q.21 A statistical model in which each of the factor has fixed effect except __________ effect is known as __________ effect model.

Q.22 A model in which some factors have fixed effect but some others have random effects, is known as __________ effect model.

Q.23 A model in which all the factors have random effect, is not known as model or __________ effect model.

Q.24 The main objective of __________ effect model is to estimate the effects of different factors involved in an experiment.

Q.25 In random effect model, the main objective to estimate the __________ among the effects of defferent factors.

Q.26 When a set of observations is classified in the form of C.R.D. then such classification is known __________ classification.

Q.27 The C. R D is based on __________ classified data.

Q.28 When a set of obervation is classified in the form R. C. B. D. then such classification is known as __________ classification.

Q.29 R. C. B. D. is the example of __________ classified data.

Q.30 The model $y_{ij} = \mu + \alpha_i + e_{ij}$, $(i = 1, 2, ...t, j = 1, 2........r)$ is suitable for __________ classified data.

Q.31 In question 30, the estimates of μ and $\propto_i$ can be worked out as

$$\mu = \frac{1}{n}\sum_{i}^{t}\sum_{j=1}^{r}\cdots\cdots\cdots \text{ and } \hat{\beta}_i = (\bar{y}_i\ldots\ldots.)$$

Q.32 Let s^2_A denotes the M. S. due to factor A in one way classified data then E (s^2A) can be expressed as

E (s^2t) =.__________ $+\phi$ (Ai, i= 1, 2, ________, t)

Q.33 In Question 32, the value of ϕ (Ai, i = 1, 2, ________, t) under Ho is equal to ________ otherwise it is a ________ quantity.

Q.34 Under Ho, the treatment M. S. (s^2_A) give an ________ estimate of ________

Q.35 If in an ANOVA test, Ho is declared significant then in place of F statistic, the ________ statistic is used to test the pairwise equality of treatment means.

Q.36 In case of R.C.B.D. the C.D. between two treatment means $(\bar{t}_{i'}, \bar{t}_i)$ with r replication can be worked out from the following expression.

$$C.D(\bar{t}i, \bar{t}i') = t_{\propto}\sqrt{\ldots\ldots\ldots\ldots\ldots}$$

10.3 Multiple Choice Question and Answers

Q.1 The basic purpose of ANOVA technique is to test the homogeneity of serveral

(a) Population means
(b) Population correlation coefficient
(c) Population variances
(d) None of the three

Q.2 The ANOVA technique was introduced first time in agricultural data by

(a) Prof. P.C. Mahalanobis
(b) Prof. R.A. Fisher
(c) Prof. P.V. Suchatme

Q.3 The ANOVA technique is now frequently used in testing the linearity of

(a) Fitted correlation coefficient
(b) Fitted regression line
(c) Fitted regression coefficient
(d) Correlation ratio n

Q.4 ANOVA technique is also used in testing the significance of

(a) Correlation ratio
(b) Correlation coefficient
(c) Regression coefficient by
(d) Homogeneity of variance

Q.5 In ANOVA model, the different components of variance should follow
(a) Multiplicative law (b) Additive law
(c) Probabilistic law (d) Large sample theory

Q.6 In ANOVA model the error term e_{ij} should be distributed as
(a) Randomly (b) Independently
(c) Normally (d) All the three

Q.7 The error term e_{ij} in ANOVA model have
(a) Mean Zero (b) Constant variance σ_e^2
(c) Same variance (d) All the three

Q.8 In ANOVA model, the constant error variance σe^2 is estimated unbiasedly by
(a) Error M. S. (b) Treatment M. S
(c) Replication M. S (d) Both (a) and (c)

Q.9 In ANOVA model, the variates y_{ij} should be ____________
(a) Homosedastic (b) Heterogeneous
(c) of equal variance (d) Both (a) and (c)

Q.10 The normality of variates y_{ij} in ANOVA model is necessary to apply the
(a) The t statistic (b) F statistic
(c) π^2 statistic (d) Z statistic

Q.11 The d. f. due to various sources of variation in ANOVA model, should follow the
(a) Probabilistic law (b) Multiplicative law
(c) Additive law (d) None

Q.12 If the Ho is true then theoretically the value of F statistic becomes
(a) Equal to 1.0 (b) Greater than 1.0
(c) Less than 1.0 (d) Equal to as ∞

Q.13 If the null hypothesis Ho in ANOVA is declared significant then theoretically the value of F statistic becomes
(a) Equal to 1.0 (b) Greator than 1.0
(c) Less than 1.96 (d) Greator than 1.96

Q.14 If s^2_t and σ_e^2 are the treatment and error M.S. in ANOVA model then the ratio s^2_t / σ_e^2 follow

(a) X^2 distribution with (t -1) d.f

(b) F distribution with (t -1) and (n -1) d. f

(c) Student's t distribution with (t -1) d. f

(d) Z distribution

Q.15 If (s^2_e σ_e^2) are the error M S with (n -t) d.f and error variance respectively then the ratio (s_e^2 / σ_e^2) will follow

(a) Student's t distribution with (t -1) d. f

(b) F distribution with (t -1) and (n -1) d. f

(c) X^2 distribution with (t -t) d. f

(d) Normal distribution with mean zero and variance unity

Q.16 If(s^2_t, σ_e^2) are treatment and error M. S. with (t-1), (n - t) d. f respectively then the ratio (s^2_t/ s^2_e) will follow.

(a) F distribution with (t -1) and (n -1) d. f

(b) χ^2 distribution with (t -1) d. f

(c) Student's t distribution with (t -1) d. f

(d) Normal distribution as N (0, 1)

Q.17 In ANOVA model, to define the F statistic, the effect due to treatment and error term must follow the

(a) Normal distribution (b) X^2 distribution

(c) Mean X^2 distribution (d) Z distribution

Q.18 F statistic is defined as the ratio of two

(a) Dependent mean X^2 statistics

(b) Independent mean X^2 statistics

(c) Two independent sample mean squares

(d) Both (a) and (c)

Q.19 The well known ANOVA technique is the technique in fact an analysis of

(a) Variance (b) Sum of square

(c) Mean sum of square (d) Both (b) and (c)

Q.20 A model in which each of the factor has fixed effect and only effect of error term is of random nature, is known as

(a) Mixed effect model (b) Random effect model
(c) Fixed effect model (d) Both (a) and (c)

Q.21 A model in which some factors have fixed effect and others have random effect, is known as

(a) Mixed effect model (b) Random effect model
(c) Fixed effect model (d) Both (b) and (c)

Q.22 A random effect model has

(a) All factors alongwith error term are of random nature
(b) Some factor are of random and others are of fixed effect
(c) All factor are of random nature except error term
(d) None of the three

Q.23 The main objective of fixed effect model is to estimate the

(a) Effect of different factors
(b) Variability of different factors
(c) Variance of different factors
(d) Both (b) and (c)

Q.24 In random effect model, the main objective is to estimate the

(a) Mean effect of different factors
(b) The variability among the effects of different factors
(c) The variance among effects due to different factors
(d) Both (b) and (c)

Q.25 When a set of observations are classified over different levels of only one factor then such type of classification, is known as

(a) Two way classification
(b) One way classification
(c) Multi way classification
(d) Both (a) and (c)

Q.26 Complete randomized design is an example of

(a) Two way classification (b) One way classification
(c) Multi way classification (d) Both (a) and (c)

Q.27 When a set of observations are classified w. r to two factors simultaneously at different levels then such classification is known as

(a) Two way classification (b) Multi way classification

(c) Both (a) and (b) (d) None of the three

Q.28 Randomised complete block design is an example of

(a) One way classification (b) Two way classification

(c) Multi way classification (d) Both (a) and (b)

Q.29 The model for one way classification is written as

(a) $y_{ij} = \beta_i + e_{ij}, \quad (i = 1, 2, ...t, j = 1, 2, ..., ni)$

(b) $y_{ij} = \mu + \beta_i + e_{ij} \quad (i = 1, 2, ...t, j = 1, 2, ..., ni)$

(c) $y_{ij} = \mu + \beta_i + \alpha_i + e_{ij} \quad (i = 1, 2.....t, j = 1, 2, ..., ni)$

(d) None of the above

Q.30 In Question 29, the β_i can be estimated through

(a) $\hat{\beta}_i = \frac{1}{n_i}\sum_{j=1}^{n} y_{ij}$

(b) $\hat{\beta}_i = \frac{1}{t}\sum_{i=1}^{t} y_{ij}$

(c) $\hat{\beta}_i = \left(\frac{1}{n_i}\sum_{j=1}^{ni} y_{ij} - \frac{1}{n}\sum_{i}^{t}\sum_{j=1}^{ni} y_{ij}\right) n_i$

(d) $\hat{\beta}_i = \frac{1}{n_i}\sum_{j=1}^{ni} y_{ij} - \mu$

Q.31 Let S_A^2 denotes the mean square due to factor A then we have

(a) $E(s^2_A) = \sigma_e^2 + \phi(Ai, i = 1, 2, ...t)$

(b) $E(s^2_A) = \sigma_e^2 - \phi(Ai, i = 1, 2, ...t)$

(c) $E(s^2_A) = \sigma_e^2 + \mu(Ai, i = 1, 2, ...t)$

(d) Both (a) and (c) Where ϕ and μ are variance like functions of (A'is,)

Q.32 When the null hypothesis Ho is true in case of one way classified data, treatment mean square gives

(a) An unbiased estimate of σ_e^2 (b) An biased estimate of σ_e^2

(c) A consistant estimate of σ_e^2 (d) None of the above

Q.33 If in Question 32, E $(s^2_A) = \sigma_e^2 + \phi(Ai,\ i = 1,\ 2, ...t)$ is true in case of under Ho, then the value of $\phi(A_{is})$ is

(a) Equal to ∞ (b) Equal to zero
(c) Greator then zero (d) Less then zero

Q.34 In Question 31, if Ho is declared significant then the value of $\phi(A_{is})$ will be

(a) Greator then zero (b) Less then zero
(c) Equal to ∞ (d) Nothing can said with surity

Q.35 In C. R. D. with r replicaion the C.D between two treatment means $(\overline{t}_{i,}\ \overline{t}_{i})$ will be given by

(a) $C\,D\,(\overline{t}_{i,}\ \overline{t}_{i'}) = t \propto \sqrt{\frac{S_e^{\ 2}}{2r}}$

(b) $C\,D\,(\overline{t}_{i,}\ \overline{t}_{i'}) = t \propto \sqrt{\frac{2\ s^2_{\ e}}{r}}$

(c) $C\,D\,(\overline{t}_{i,}\ \overline{t}_{i'}) = t \propto \sqrt{\frac{r\,s_e^{\ 2}}{2}}$

(d) $C\,D\,(\overline{t}_{i,}\ \overline{t}_{i'}) = t \propto \sqrt{\frac{s_e^{\ 2}}{2r}}$

10.4 Key Answers

10.4.1 True / False

1.	False	2.	True	3.	False
4.	True	5.	True	6.	True
7.	True	8.	True	9.	True
10.	True	11.	True	12.	True
13.	False	14.	True	15.	True
16.	True	17.	True	18.	False
19.	True	20.	True	21.	True
22.	True	23.	False	24.	True

25.	True	26.	True	27.	False
28.	True	29.	True	30.	True
31.	True	32.	True	33.	True
34.	True	35.	True	36.	True
37.	True				

10.4.2 Fill in the Blanks

1. Variance, Variance
2. Homogeneity
3. R.A. Fisher
4. Regression
5. Correlations
6. Random
7. Additive
8. Independently Normally
9. Zero, variance
10. Unbiased
11. ANOVA, Error
12. Normally, F
13. Additive
14. Unity, Unity
15. Independently, Chi-square
16. Normally
17. Independent
18. Not Add
19. Sum of square
20. Error
21. Error, Fixed
22. Mixed
23. Fixed, Mixed
24. Fixed
25. Variability
26. One Way

27. One Way
28. Two Way
29. Two Way
30. One Way
31. $y_{ij}, (\overline{y}_i - \hat{\mu})$
32. σ_e^2
33. Unity, Positive
34. Unbiased , Error Variance σ_e^2
35. Student' t
36. $\sqrt{\frac{2\,s_e^2}{r}}$

10.4.3 Multiple Choice Question and Answers

1.	a	2.	b	3.	b
4.	a	5.	b	6.	d
7.	d	8.	a	9.	d
10.	b	11.	c	12.	a
13.	b	14.	a	15.	c
16.	b	17.	a	18.	b
19.	d	20.	c	21.	a
22.	a	23.	a	24.	d
25.	a	26.	b	27.	a
28.	b	29.	b	30.	d
31.	d	32.	a	33.	b
34.	a	35.	b		

11

C.R.D., R.C.B.D. Latin Square Design

11.1 True / False

Q.1 We can apply the change of origin or change of scale or change of origin and scale both to the original data in case of applying ANOVA technique to make the analysis easy and speedy.

Q.2 The conclusions drawn from ANOVA based on transformed data or original data are different.

Q.3 The statistical model for two way classified observations with one observation per cell is written as

$y_{ij} = \mu + \alpha_i + \beta_i + r_{ij} + e_{ij}$

where r_{ij} is the interaction between ith level of A and jth level of B.

Q.4 If the interaction $r_{ij} = (A i\, Bj) \neq \alpha_i + \beta j$ then interaction r_{ij} will not present in two way classified data

Q.5 In Question 3, an unbiased estimate of α_i and β_j, can be obtained from the relation.

$\hat{\alpha}_i = \overline{y}i - \overline{y}..$ and $\hat{\beta}_j = \overline{y}_j - \overline{y}..$

where $\overline{y}..$ is the grand mean of y_{ij}

Q.6 The functional forms of M. S. (A) and M. S. (B) will be written as

M. S. (A) = $\sigma_e^{\,2} + \phi(\alpha_i^2)$ and M. S. (B) = $\sigma_e^{\,2} + \mu(\beta_j^2)$

where $\phi(\alpha i^2)$ and $\mu(\beta_j^{\,2})$ will the variance like terms.

Q.7 If the null hypotheses for both the factors A and B are true then $\phi(\alpha_i^2)$ and $\mu(\beta_j^2)$ will be equal to zero.

Q.8 If the null hypotheses Ho1 $(\alpha_i) = 0$ and Ho2 $(\beta_j) = 0$ are non-significant then both E (M. S. A) and E (M. S. B), will give the unbiased estimates of $\sigma_e^{\,2}$.

Q.9 Out of three basic designs C.R.D., R.C.B.D. and L.S.D., the simplest design is L.S.D.

Q.10 To apply C.R.D. the experiment material should not be non-homogeneous.

Q.11 Two basic principles followed by C. R. D. are the randomisation and local control.

Q.12 The C.R.D. permits the complete flexibility regarding the numbers of treatment and replication to be followed in it.

Q.13 The principle of randomisation assures that the extraneous factors (error terms) do not continuously influence one treatment.

Q.14 C.R.D. provide the maximum number of d. f to estimate the error variance σ_e^2.

Q.15 The sensitivity or precision of the experiment for small experiment (i e experiment with small number of treatment) is permitted in C.R.D.

Q.16 In methodological studies like physics, chemistry, cookery, in chemical and biological experiment in some green house, an experimental design used is the C.R.D.

Q.17 For two treatments (i, j) being replicated (ri, rj) times in C.R.D., the S.E. of difference $(\overline{x}_i - \overline{x}_j)$ will be estimated as-

$$\hat{S}.E\,(\overline{x}_i - \overline{x}_j) = \sqrt{s_e^2\left(\frac{1}{ri} + \frac{1}{rj}\right)} \text{ with } \hat{\sigma}_e^{\,2} = S_e^{\,2}$$

11.2 Randomised Complete Block Design (R.C.B.D.)

Q.18 All the three basis principles of design of experiment enumerated by Prof. R.A. Fisher are not followed in R.C.B.D.

Q.19 In R.C.B.D., each treatment is not replicated equal numbers of time.

Q.20 Analysis of R.C.B.D. is analogous to the analysis of two way classified data with one observation per cell.

Q.21 In R. C. B. D, the randomisation is done separately within each block.

Q.22 In R.C.B.D., the object is to keep the experimental error within each block as small as possible.

Q.23 Let one observation (x), in treatment 1, in block 1, in R.C.B.D. is missing then the missing value x can be estimated from the expression.

$$x = \frac{(r T' + r B' - G')}{(r-1)(t-1)}$$

where T & B are the total of treatment 1 and block 1 respectively and G' is the grand total of available values.

Q.24 In case of one observation missing in R.C.B.D., then error and total d. f. can be worked out as-

Error d. f. = (r - 1) (t - 1) -1 and total d. f = (rt - 1) -1

Q.25 If R.C.B.D. is conducted as C.R.D. i e blocking is not considered for analysis, then error M.S.E (S'^{2}_{e}) of C.R.D. will be worked out from-

$$s'^{2}_{e}\ (\text{CRD}) = \frac{n_r s^2_r + (n_r + n_e)s^2_e}{(n_r + n_t + n_e)}$$

where (n_r, n_t, ne) are the d. f. for replication, treatment and error in R.C.B.D., respectively and s'^{2}_{e} be error M.S. in R.C.B.D.

Q.26 In Question 25, the efficiency of R.C.B.D. over C.R.D. is defined as

$$\text{Efficiency (R.C.B.D./C.R.D.)} = \frac{s'^{2}_{e}(\text{C.R.D.})}{s^{2}_{e}(\text{R.C.B.D.})} \times 100$$

Q.27 If more than two observations are missing in R.C.B.D., then these missing observations can be estimated using the technique known as "Bartlett's missing plot technique"

Q.28 In case of m observations missing in R.C.B.D., the error d. f. will be worked out as

Error (d. f.) = (r-1) (t-1) - m

Q.29 If the null hypothesis Ho for treatments and blocks are found non-significant then treatment M.S. and block M.S., give unbiased estimate of σ_e^2.

Q.30 If some extra replications are required for two treatment t_1 & t_2 in R.C.B.D. they are replicated (r_1 & r_2) time respectively then estimate of S.E. (t_1-t_2) will be given as

$$\widehat{\text{S.E.}}\ (\overline{t}_1 - \overline{t}_2) = S_e\sqrt{\frac{1}{r_1} + \frac{1}{r_2}}$$

where S_e^2 is the error M.S.

11.3 Latin Square Design (L.S.D.)

Q.31 In agricultural field experiment where fertility contour map is not known then L.S.D. is suitable to apply.

Q.32 In latins square design, the heterogeneity of experimental material is not controlled simultaneously in two directions at right angle.

Q.33 In industry and agricultural field experiment, the L.S.D. is suitable to be applied.

Q.34 In L.S.D. the number of treatments and replications are not same.

Q.35 In case of m x m, L.S.D., the number of experimental unit is equal to m^3.

Q.36 In L.S.D., the m treatments are allocated at random to rows and columns in such a way that every treatment occurs once and only once in each row and each column.

Q.37 In L.S.D., the m^2 experimental unit will be arranged in a square.

Q.38 In case A, B, C, D, being four treatments, the following L S D arrangement is suitable.

A	B	C	D
B	C	D	A
C	B	D	A
D	A	B	C

Q.39 The number of basic principles of design of experiment followed in L.S.D. is equal to 3.

Q.40 L.S.D. is an incomplete three way layout.

Q.41 In an incomplete three way layout, only m^2 experimental units are utilised.

Q.42 In a complete three way layout m^3 experimental units are utilised.

Q.43 In L.S.D. the suitable approximate number of treatments is between (5 to 10).

Q.44 In m x m, L.S.D., the error d. f. is equal to (m - 1) (m - 2)

Q.45 The linear model for m x m, L.S.D. is written as

$$y_{ijk} = \mu + \alpha_i + \beta_j + t_k + e_{ijk},\ (i, j, k = 1, 2........m)$$

Q.46 In an incomplete 3 way lay out (ie L.S.D.) the three factors of variation in observations are, the row, column and treatment respectively.

Q.47 In 7 x 7, L.S.D., the error d.f. is equal to 30

Q.48 With the help of contrast, one can estimate the linear effect.

Q.49 The error S.S. in R.C.B.D. as compared to C.R.D. using the same experimental is less.

Q.50 Another name of critical difference (C.D.) is the least significant difference (L.S.D.).

Q.51 The plan of an experiment which controls all factors as for as possible except the treatment effect, is known as design of experiment.

Q.52 For the validity of probability statements about the treatment difference in an experimental design, the suitable principle (technique) used is the - randomisation.

Q.53 S.S. due to contrast $Y = -T_1 - T_2 + T_3 + T_4$ Where each treatment total (Ti) is the sum of 4 observation, is equal to $(Y^2/4^2)$

Q.54 In L.S.D. in which the letters occur in alphabetical order is called as - "Reduced Latin Square"

11.4 Incomplete Block Design (I.B.D.)

Q.55 A block design in which each block does not contain a complete set of treatments under study, is known as "Incomplete Block Design"

Q.56 In an I.B.D., the number of treatments under study is less than the block size.

11.5 Notations

We consider the following notations to study the theory of incomplete block design.

1. V : The number of treatments under investigation
2. b : The number of blocks in design
3. K : The size of each block
4. r : The number of replication for each treatment
5. λ : The number of blocks in which each pair of treatments occurs together.

Q.57 In a balanced incomplete block design, each pairs of treatments occurs together in λ blocks.

Q.58 The integers v, b, r, k and λ are known as the parameters of B.I.B.D.

Q.59 The two parametric relations (i) vr =bk and (ii) λ(V-1) = r (k -1) are two necessary conditions out of three conditions for the existence of B.I.B.D.

Q.60 The inequality b $\geq$ k is known as fisher inequality

Q.61 The parametric relations are only the necessary conditions for the existence of B.I.B.D.

Q.62 If for any B.I.B.D., λ =r then B.I.B.D. reduces to a complete block design.

Q.63 If for any B.I.B.D., λ = r then block size k will be equal to the number of treatment.

Q.64 If any two blocks in B. I. B. D intersect at a constant number of treatment (say λ) then B.I.B.D. is said to be symmetrical design.

Q.65 For a B.I.B.D., there exists a complimentary design which is also B.I.B.D.

Q.66 For a given symmetrical B.I.B.D. (say D), there exists is residual design say D*.

Q.67 The residual design D* is also a B.I.B.D.

Q.68 An incomplete block design is said to be connected, if all the elementary contrasts of treatments are estimable otherwise it is called disconnected.

Q.69 If in an incomplete block design, all the elementary contrasts are estimated with the same precision then the design is said to be balanced.

Q.71 The efficiency factor of B.I.B.D., is greater than unity.

11.6 Fill in the Blanks

Q.1 We can apply the change origin or scale or origin & scale both to the original data before applying __________ technique

Q.2 The conclusion drawn from ANOVA based on transformed data or original data are __________

Q.3 The statistical model for two way classified data with one observation per cell can be written if interaction is not present.

y_{ij} = __________

Q.4 If the interaction $r_{ij} \neq \alpha_i + \beta_j$ then the effects of two factors A and B will be __________ in two way classified data.

Q.5 The functional form for M.S.A. and M.S.B. can be written as

M.S.A. = σ_e^2 + __________, M.S.B. = σ_e^2 + __________

Q.6 If M.S.A. $= \sigma_e^2 + \phi(\alpha i^2)$ and M.S.B. $= \sigma_e^2 + \mu(\beta_i^2)$, then both $\phi(\alpha_i^2)$ and $\mu(\beta_i^2)$ will be __________ like terms.

Q.7 If the null hypotheses Ho for both the effects α_i and β_j are declared non significant then value of $\phi(\alpha^2_i)$ and $\mu(\beta_j^2)$ both will be equal to __________

Q.8 In case of both hypotheses Ho_1 $(\alpha i) = 0$ and Ho_2 $(\beta_j) = 0$ being declared non significant then both E (M.S.A.) and E (M.S.B.) will give __________ estimates of σ_e^2

Q.9 Out of three basis designs C.R.D., R.C.B.D. and L.S.D. the simplest design is __________

Q.10 To apply C.R.D., the experimental material must be of __________ nature.

Q.11 Two basic principles followed in C.R.D. are the __________ and __________

Q.12 The C.R.D. permits to complete flexibility in deciding the number of __________ and __________ to be included in it.

Q.13 The principle of __________ assures that extraneous factors do not influence continuously the same treatment.

Q.14 C.R.D. provides the __________ number of d. f. to estimate the error variance σ_e^2

Q.15 The sensitivity or precision of the experiment for small experiment is permitted in __________

Q.16 In methodological studies like physics, chemistry, cookery, in chemical and biological experiments in some green house, experimental design used is the __________

Q.17 With two treatments (i, j) being replicated (r_i, rj) times respectively in C. R. D., the S. E. $(\overline{x}_i - \overline{xj})$ will be estimated as $S.E(\overline{x}_i - \overline{xj}) = \sqrt{Se^2(........)}$, with $E(s_e^2) = \sigma_e^2$

11.7 Randomised Complete Block Design (R.C.B.D. or R.B.D.)

Q.18 The number of basic principle of design of experiments followed in R.C.B.D. is equal to ________

Q.19 In R.C.B.D. each treatment is replicated ________ number of times.

Q.20 Analysis of R.C.B.D. is analogous to the analysis of ________ classified data with one observation per cell.

Q.21 In R.C.B.D., the randomisation is done separately and independently ________ each block.

Q.22 In R.C.B.D., the object is to keep ________ error as small as possible ________ each block.

Q.23 In case of two observations missing in R.C.B.D. with (r, t) number of (replication and treatment) we have error d.f = (r - 1) ( ) ________, Total d. f = (r_t - 1) ________

Q.24 For R.C.B.D. to be conducted as C.R.D. with the same experimental material the error M.S. (C.R.D.) will be given as

$$\text{Error MC (C.R.D.)} = \frac{n_r\, s^2 r + (........)\, s^2 e}{(.........)}$$

where (n_r. n_t. n_e) are d. f. for replication treatment and error respectively and s^2e is error M.S. in R.C.B.D.

Q.25 $$\text{Efficiency (R.C.B.D./C.R.D.)} = \frac{\text{Error M.S.(........)}}{\text{Error M.S.(........)}} \times 100$$

Q.26 For missing three or more observations in R.C.B.D., "________ missing plot technique" will be used to estimate them.

Q.27 For m observations missing in R.C.B.D., we have error (d.f) = (r -1) (t - 1) ________

Q.28 In case of null hypotheses for (treatment & block) being declared non-significant, the M.S. (treatment and block) give ________ estimate of σ_e^2

Q.29 In case of Ho (treatment & block) being found "________", the M.S. (Treatment & block) give unbiased estimate of σ_e^2

Q.30 In R.C.B.D., two treatment (t_1, & t_2) being replicated (r_1, & r_2) times with Error M.S. (s^2_e), we have S. E. (t_1-t_2)

$$\widehat{\text{S.E.}}(\overline{t}_1 - \overline{t}_2) = {}^{Se}\sqrt{(.......)}$$

11.8 Latin Square Design (L.S.D.)

Q.31 In agricultural field experiment where fertility contour map is not known, then suitable design applicable is _________

Q.32 In L.S.D. the principle of local control is applied in two _________ directions.

Q.33 In mxm L.S.D., the number of replication and treatment will be equal to _________

Q.34 In mxm, L.S.D. the number of experimental units is equal to _________

Q.35 In L.S.D. each treatment must, occur once and _________ in each row and each _________

Q.36 In L.S.D. every treatment must occur _________ and only _________ in each _________ and each column.

Q.37 In mxm L.S.D. the m^2 _________ units will be arranged in a square.

Q.38 In 4x4, L.S.D., the suitable arrangement is as-

A B C D

B C (- -)

C D (- -)

D A (- -)

Q.39 The number of basic principles of design of experiment followed in L.S.D., is equal to _________

Q.40 L.S.D. is an incomplete _________ layout.

Q.41 In an incomplete three way layout (i. e. m x m), the number of experimental units is equal to _________

Q.42 In a complete three way (m x m x m), the number of experimental units will be equal to _________

Q.43 In L.S.D. the suitable approximate number of treatments is from (_________ to _________)

Q.44 In m x m L.S.D. error d. f. is equal to _________

Q.45 The linear model for mxm, L.S.D. is written as

$$y_{ijk} = \mu + ... + \beta_j + t_k + ..., (i = j = k = 1, 2, ..., m)$$

Q.46 In an incomplete three way layout (i.e L.S.D.) the three factors responsible for variation in observations are _________, _________ and _________ respectively.

Q.47 In 7x7, L.S.D., the error d. f. in equal to _________

Q.48 With the help of contrasts, one can estimate the _________ effect.

Q.49 The Error S.S. in R.C.B.D. as compared to _________ using the same experimental material is less.

Q.50 Another name for "Critical Difference" is _________

Q.51 The error M.S. in L.S.D. is less than that in either _________ or _________, using the same experimental material.

Q.52 The plan of an experiment which controls all the factors as for as possible except treatment under investigation, is known as _________ of experiment.

Q.53 For the validity of probability statements about the treatment differences in an experimental design, the suitable principle of design is the _________

Q.54 The S.S. due to contrast $Y = T_1 - T_2 + T_3 - T_4$ with each treatment total (Ti,) being the sum of r observation is equal to _________

Q.55 In L.S.D. in which letter occur in alphabatical order, is called as "_________"

11.9 Incomplete Block Design (I.B.D.)

Q.56 A block design in which each block does not contain the complete treatment, is known as an _________ block design.

Q.57 In an I.B.D., the number of treatment is _________ than the block size.

Q.58 In a B.I.B.D., each pair of treatment occurs together in l _________

Q.59 The integers (v, b, r, k & λ) are known as _________ of B.I.B.D.

Q.60 The parametric relations (i) vr = rk and (ii) λ (v - 1) = r (k - 1) are the two _________ out of three for the existence of B.I.B.D.

Q.61 The inequality $b \geq v$ is known as _________ inequality.

Q.62 The parametric relations are the only _________ for the existence of B.I.B.D.

Q.63 For any B.I.B.D., if λ = r then B.I.B.D. refuses to a _________ design.

Q.64 For any B.I.B.D., if = r then block size k will be equal to the number of _________ under investigation.

Q.65 If any two blocks in B.I.B.D., intersect at a constant number of treatment (say λ) then B.I.B.D. is said to be a _________ design

Q.66 For a B.I.B.D., there exists a complementary design which is also _________

Q.67 A residual design D* is also a _________

Q.68 An incomplete block design is said to be _________ if all the _________ contrasts of treatment are estimable otherwise it is called_________

Q.69 For a given symmetrical B.I.B.D. (say D) their exists a residual _________ say D*

Q.70 An incomplete block design is said to be _________ if all the _________ contrasts of treatment are estimable with the equal _________

Q.71 The designs R.C.B.D. and L.S.D., both the orthogonal, _________ and _________

Q.72 The efficiency factor of B.I.B.D. is less then _________

Q.73 The _________ factor of B.I.B.D. is _________ than, unity.

Q.74 The design R.C.B.D. and L.S.D. both are orthogonal, _________ and balanced.

Q.75 The designs R.C.B.D. and L.S.D., both are orthogonal, connected and _________

11.10 Multiple Choice Questions and Answers

Q.1 We can transform the original data before applying the ANOVA technique using change of-

(a) Origin (b) Scale
(c) Origin and Scale both (d) All the three

Q.2 The result drawn from ANOVA based on transformed data is -

(a) Same as based on original data
(b) Different from original data
(c) More efficient than based on original data
(d) Less efficient than based on original data

Q.3 For two factors A and B being dependent, the statistical model with one observation per cell can be wirtten as

(a) $y_{ij} = \mu + \alpha_{ij} + r_{ij} + e_{ij}$ (b) $y_{ij} = \mu + \beta_i + r_{ij} + e_{ij}$

(c) $y_{ij} = \mu + \alpha_\ell + \beta_j + r_{ij} + e_{ij}$ (d) $y_{ij} = \mu + \alpha_\ell + \beta_j + e_{ij}$

Q.4 For the interaction r_{ij} = Ai Bj to be present in the two way classified data, we have.

(a) $\alpha_i + \beta_j \neq r_{ij}$ (b) $\alpha_i + \beta_j = r_{ij}$

(b) $\alpha_i + \beta_j \geq r_{ij}$ (d) $\alpha_i + \beta_j \leq r_{ij}$

Q.5 In Q.3, the effect α_i can be estimated as

(a) $\hat{\alpha}_i = \overline{y}_i - \overline{y}..$ (b) $\hat{\alpha}_i = \overline{y}_{ij} - \overline{y}..$

(c) $\hat{\alpha}_i = \overline{y}_{ij} - \overline{y}..$ (d) $\hat{\alpha}_i = \overline{y}_{ij} - \overline{y}_i$

Q.6 In Q.3, the block effect B_j can be estimated as.

(a) $\hat{\beta}_j = \overline{y}_i - \overline{y}..$ (b) $\hat{\beta}_j = \overline{y}_j - \overline{y}..$

(c) $\hat{\beta}_j = \overline{y}_{ij} - \overline{y}_{\cdot j}$ (d) $\hat{\beta}_j = \overline{y}_{ij} - \overline{y}_{i\cdot}$

Q.7 The functional form for M. S. A can be written as

(a) M. S. A $= \sigma_e^2 + \phi(\alpha_i^2) + \mu(\beta_j^2)$

(b) M. S. A $= \sigma_e^2 + \phi(\alpha_i^2)$

(c) M. S. A $= \sigma_e^2 + \mu(\beta_j^2)$

(d) M. S. A $\neq \sigma_e^2 + \phi(\alpha_j^2)$

Q.8 The functions form for M. S. B. can be written as

(a) M. S. A $= \sigma_e^2 + \phi(\alpha_i^2)$

(b) M. S. A $= \sigma_e^2 + \phi(\alpha_i^2) + \mu(\beta_j^2)$

(c) M. S. A $= \sigma_e^2 + \mu(\beta_j^2)$

(d) M. S. A $\leq \sigma_e^2 + \phi(\alpha_j^2) + \mu(\beta_j^2)$

Q.9 If the null hypothesis Ho: $\alpha_i = 0$, is true then $\phi(\alpha_i^2)$ will be:

(a) $\phi(\alpha_i^2) += 1$ (b) $\phi(\alpha_i^2) \neq 1$

(c) $\phi(\alpha_i^2) = 0$ (d) $\phi(\alpha_i^2) \neq 0$

Q.10 If Ho $(\alpha_i^2 = 0)$, is found to be nonsignificant then E. (M.S.A.) will -

(a) Give an unbiased estimate of σ_e^2

(b) Not give an unbiased estimate of σ_e^2

(c) Give a biased estimate of σ_e^2

(d) Not give a biased estimate of σ_e^2

Q.11 If in two way classified data, Ho: $B_j = 0$ is found non significant then E. (M.S.B.) will -

(a) Give an unbiased estimate of σ_e^2

(b) Not give an unbiased estimate of σ_e^2

(c) Not give a biased estimate of σ_e^2

(d) Give a biased estimate of σ_e^2

Q.12 Out of three designs, C.R.D., R.C.B.D. and L.S.D., the simplast design is-

(a) L.S.D. (b) R.C.B.D.

(c) C.R.D. (d) None

Q.13 To apply C.R.D. the experimental material should -

(a) Not be homogeneous (b) Be homogeneous

(c) Not be non-homogeneous (d) Both (b) and (c)

Q.14 Two basic principles followed C.R.D. are

(a) Randomisation and local control

(b) Randomisation and Replication

(c) Randomisation and local control

(d) Randomisation and Regularity

Q.15 C.R.D. permits the complete flexibiltiy in-

(a) No and randomisation replications

(b) No of treatment and replicalion

(c) No of treatment and randomisation

(d) All the above three

Q.16 The principle which assures that extraneous factors do not influence continuously one factor is -

(a) Replication (b) Randomisation

(c) Local control (d) Regularity

Q.17 The design which provides maximum number of possible d. f to estimate is the-

(a) R.C.B.D. (b) C.R.D.

(c) L.S.D. (d) B.I.B.D.

Q.18 The sensibility or precision of an experiment in case of a small experiment is permitted in-

(a) L.S.D. (b) C.R.D.

(c) R.C.B.D. (d) Both (a) and (c)

Q.19 In methodological studies like physics chemistry cookery, in same green house experiment the design applicable is-

(a) L.S.D. (b) R.C.B.D.

(c) C.R.D. (d) B.I.B.D.

Q.20 In C.R.D. the two treatment (i, j) being replicated (r_i, rj) times, the S.E. $(\overline{x}i - \overline{x},)$ will be estimated by-

(a) $\sqrt[2Se]{\left(\frac{1}{ri}+\frac{1}{rj}\right)}$ (b) $\sqrt{S^2{}_e\left(\frac{1}{ri}+\frac{1}{rj}\right)}$

(c) $\sqrt[S^2_e]{\left(\frac{1}{ri}+\frac{1}{rj}\right)}$ (d) $\sqrt[S^2_e]{\left(\frac{1}{ri}+\frac{1}{rj}\right)}$

11.11 Randomisation Complete Block Design (C.R.B.D.)

Q.21 The number of basic principles followed in R.C.B.B.D. is equal to-

(a) 1 (b) 2

(c) 3 (d) 5

Q.22 Analysis of R.C.B.D. is analogous to the analysis of -

(a) One way classified data

(b) Two way classified data with one observations per cell.

(c) Incomplete three way classified data

(d) None of the above

Q.23 In R.C.B.D., the randomisation is done-

(a) Separately within each block

(b) Independently within each block

(c) Not independently within each block

(d) As (a) and (b) both

Q.24 In R.C.B.D. the object is to keep the experimental error-

(a) Within each block as high as possible

(b) Within each block as small as possible

(c) Between two adjacent blocks as small as possible

(d) Both (b) and (c)

Q.25 In case of m observation missing in R.C.B.D., with the nubmer of treatments and replications as (t, r) respectively the error d. f. will be equal to-

(a) (r - 1) (t - 1) + m (b) (r - 1) (t - 1) - m

(c) (rt - 1 - m) (d) (rt - 1) - (m - 1)

Q.26 If R.C.B.D. is conducted as R.C.D. with the same experimental material then M.S. (C.R.D.) will be worked out from the expression.

(a) $\frac{n_r s_t^2 + (n_r + n_e) s_e^2}{(n_r + n_t + n_e)}$ (b) $\frac{n_r s_r^2 + (n_t + n_e) s_e^2}{(n_r + n_t + n_e)}$

(c) $\frac{n_r s_r^2 + (n_r + n_e) s_e^2}{(n_r + n_t + n_e)}$ (d) $\frac{n_t s_r^2 + (n_r + n_e) s_e^2}{(n_r + n_t + n_e)}$

Where (nr, nt, ne, s_e^2) have usual meaning.

Q.27 For more than two observations missing in R.C.B.D., to estimate them, the technique used is known as-

(a) Bartllet's missing plot technique

(b) Fisher's missing plot technique

(c) Kemthorn's missing plot technique

(d) Cochran and cox's missing plot technique

Q.28 In R.C.B.D., the treatment M.S. will give an unbiased estimate of σ_e^2 when-

(a) Block effect is declared non-significant

(b) Treatment effect is declared non-significant

(c) Block effect is declared significant

(d) Treatment effect is declared significant

Q.29 In R.C.B.D. the block M.S. will provide an unbiased estimate of σ_e^2 when-

(a) Block effect is declared significant
(b) Treatment effect is declared non-significant
(c) Treatment effect is declared significant
(d) Block effect is declared non significant

Q.30 If in case of some extra replication in R.C.B.D., the two treatments (t_1 & t_2) are replicated (r_1 & r_2) time with error M.S. as s^2_e then S. E. $(\bar{t}_1 - \bar{t}_2)$ will be estimated by-

(a) $\sqrt[S^2e]{\left(\frac{1}{r_1}+\frac{1}{r_2}\right)}$ (b) $\sqrt[2Se]{\left(\frac{1}{r_1}+\frac{1}{r_2}\right)}$

(c) $\sqrt[2Se^2]{\left(\frac{1}{r_1}+\frac{1}{r_2}\right)}$ (d) $\sqrt{2s_e^2\left(\frac{1}{r_1}+\frac{1}{r_2}\right)}$

11.12 Latin Square Design (L.S.D.)

Q.31 In agricultural field experiment where fertility contour map is not known then a suitable design applicable is the -

(a) R.C.B.D. (b) C.R.D.
(c) L.S.D. (d) B.I.B.D.

Q.32 In L.S.D., the experimental material of heterogeneous nature is controlled simultmeously in-

(a) Two parallel directions
(b) Two perpendiculer directions
(c) Incomplete three way classifications
(d) Both (b) and (c)

Q.33 In a m x m, L.S.D. the number of experimental units is equal to-

(a) m (b) m^2
(c) m^3 (d) (m - 1) (m -2)

Q.34 In case of four treatments being A, B, C, D for L.S.D. the following layout plan is correct-

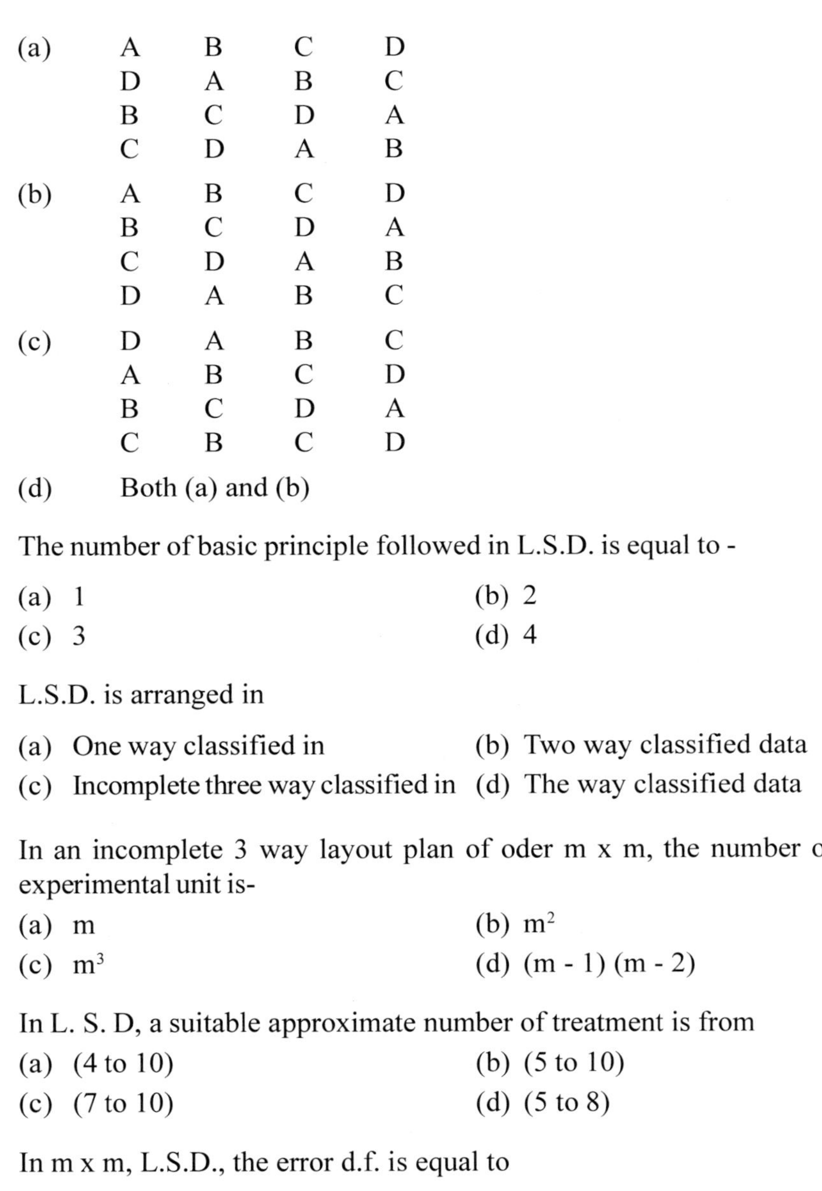

(a)

A B C D
D A B C
B C D A
C D A B

(b)

A B C D
B C D A
C D A B
D A B C

(c)

D A B C
A B C D
B C D A
C B C D

(d) Both (a) and (b)

Q.35 The number of basic principle followed in L.S.D. is equal to -

(a) 1
(b) 2
(c) 3
(d) 4

Q.36 L.S.D. is arranged in

(a) One way classified in
(b) Two way classified data
(c) Incomplete three way classified in
(d) The way classified data

Q.37 In an incomplete 3 way layout plan of oder m x m, the number of experimental unit is-

(a) m
(b) m^2
(c) m^3
(d) (m - 1) (m - 2)

Q.38 In L. S. D, a suitable approximate number of treatment is from

(a) (4 to 10)
(b) (5 to 10)
(c) (7 to 10)
(d) (5 to 8)

Q.39 In m x m, L.S.D., the error d.f. is equal to

(a) (m - 1) (m + 1)
(b) m (m - 1)
(c) (m - 1) (m - 2)
(d) m (m - 2)

Q.40 The suitable linear model for m x m, L.S.D. is

(a) $y_{ijk} = \mu + \alpha_i + \beta_j + e_{ijk}$

(b) $y_{ijk} = \mu + \alpha_i + t_k + e_{ijk}$ (i =j = k = 1, 2.......m)

(c) $y_{ijk} = \mu + \alpha_i + \beta_j + t_k + e_{ij}$

(d) $y_{ijk} = \mu + \alpha_i + \beta_j + t_k + e_{ijk}$

Q.41 In an incomplete 3 way layout experiment, the three factors responsible for variation in observation are -

(a) Row, Column and Blocks
(b) Row, Column and Replication
(c) Row, Column and Treatment
(d) Row, Treatment and Blocks

Q.42 In 7 x 7 , L.S.D., the error d. f. is equal to

(a) 42 (b) 36
(c) 35 (d) 30

Q.43 Error S.S. in L.S.D. using the some experimental material will be-

(a) Less than that in C.R.D.
(b) Same as in C.R.D. and R.C.B.D.
(c) Less than that in R.C.B.D.
(d) Both (a) and (c)

Q.44 The error M.S. in L.S.D. using the same experimental material will be-

(a) Less then that in R.C.B.D.
(b) Same as in R.C.B.D. and R.C.R.D.
(c) Less than that in C.R.D. and R.C.B.D.
(d) Greater than that in R.C.B.D.

Q.45 For the validity of probability statements about the treatment differences in an experimental design, the suitable principles used is-
(a) Local control (b) Randomisation
(c) Replication (d) None

Q.46 S.S. due to contrast $Y = T_1 - T_2 + T_4 - T_3$ where each treatment total (Ti) is the sum of 5 observations is equal to-

(a) $\frac{Y^2}{16}$ (b) $\frac{Y^2}{25}$

(b) $\frac{Y^2}{20}$ (d) $\frac{Y^2}{15}$

11.13 Incomplete Block Design (I.B.D.)

Q.47 In an I.B.D., the number of treatment under study is-
(a) Less than block size (b) Equal to block size
(c) Greater than block size (d) Equal to number of blocks

Q.48 A set of parameters of B.I.B.D. is -
(a) v, b, r, λ (b) v, b, k, λ
(c) v, b, r, k, λ (d) v, b, λ, k

Q.49 In B.I.B.D., each pair of treatment (say t_i, t_j, $i \neq j$) occurs together in-
(a) λ number of block (b) (λ-1) number of block
(c) (λ+1) number of block (d) None of the three

Q.50 The inequality $b \geq k$ is known as-
(a) Kemthorn's in-equality (b) Fisher's in equality
(c) Cochran 8 cox inequality (d) Neyman's inequality

Q.51 For the existance of BIBD, the parametric relations are-
(a) Only sufficient conditions
(b) Only necessary conditions
(c) Necessary and sufficient conditions
(d) Not necessary conditions

Q.52 In case of λ = r in B.I.B.D., then B.I.B.D. reduces to -
(a) Complete block design (b) Randomised block design
(c) Complete Randomised design (d) Both (a) and (b)

Q.53 For any B.I.B.D., if λ=r, k will be :
(a) Equal to number of blocks
(b) Equal to number of replications
(c) Equal to number of treatment
(d) Both (a) and (b)

Q.54 If in a B.I.B.D., any two blocks intersect at a constant number of treatment (say λ), then design is said to be-
(a) Assymmetrical (b) Symmetrical
(c) Complete design (d) Unconfounded design

Q.55 A given I.B.D., is also a ....

(a) Balanced Design (b) Connected Design
(c) Confounded Design (d) Both (a) & (b)

Q.56 For a given B.I.B.D., there exists-

(a) Complementary design (b) Residual design
(c) Both (a) and (b) (d) None

Q.57 The designs R.C.B.D. and L.S.B., both are

(a) Balanced designs (b) Connected designs
(c) Orthogonal designs (d) All the three

Q.58 The efficiency factor of a B.I.B.D. is -

(a) Less than 1.0 (b) Greater then 1.0
(c) Greater than 2.0 (d) None of three

11.14 Key Answers

11.14.1 True / False

1.	True	2.	False	3.	True
4.	False	5.	True	6.	True
7.	True	8.	True	9.	False
10.	True	11.	False	12.	True
13.	True	14.	True	15.	True
16.	True	17.	True	18.	False
19.	False	20.	True	21.	True
22.	True	23.	True	24.	True
25.	True	26.	True	27.	True
28.	True	29.	True	30.	True
31.	True	32.	False	33.	True
34	False	35.	False	36.	True
37.	True	38.	False	39.	True
40.	True	41.	True	42.	True
43.	True	44.	True	45.	True

46.	True	47.	True	48.	True
49.	True	50.	True	51.	True
52.	True	53.	True	54.	True
55.	True	56.	True	57.	False
58.	True	59.	True	60.	True
61.	True	62.	True	63.	True
64.	True	65.	True	66.	True
67.	True	68.	True	69.	True
70.	True	71.	False	72.	False

11.14.2 Fill in the Blanks

1. ANOVA
2. Same
3. $y_{ij} = \mu + \alpha_i + \beta_j + e_{ij}$
4. Dependent
5. $\phi(\alpha_i^2)\eta(\beta_j^2)$
6. Variance
7. Zero
8. Unbiased
9. C.R.D.
10. Homogeneous
11. Randomisation and local control
12. Replication, Treatment
13. Randomisation
14. Maximum
15. C.R.D.
16. C.R.D.
17. $\left(\frac{1}{r_i} + \frac{1}{r_j}\right)$
18. Three

19. Equal or same
20. Two way
21. Within
22. Experimental, within
23. (t -1) -2, -2
24. $\frac{(n_r + n_e)}{(n_r + n_t + n_e)}$
25. C.R.D./ C.R.B.D.
26. Bartllet's
27. - m
28. Unbiased
29. Non-significant
30. $\left(\frac{1}{r_1} + \frac{1}{r_2}\right)$
31. L.S.D.
32. Perpendicular
33. m
34. m^2
35. Only once, column
36. Once, once Row
37. Experimental units
38. (D, A), (A, B) (B, C)
39. Three
40. Three way
41. m^2
42. m^3
43. (5 to 10)
44. (m - 1) (m - 2)
45. $\alpha_i \, e_{ijk}$
46. Row, column Treatment

47. 30
48. Linear
49. C.R.D.
50. Least significant difference
51. C.R.D., R.C.B.D.
52. Design
53. Randomisation
54. Randomisation
55. $\dfrac{Y^2}{r^2}$
56. Reduced Latin square
57. Incomplete
58. Greater
59. Blocks
60. Parameters
61. Necessary conditions
62. Fisher's
63. Necessary conditions.
64. Randomised complete Block
65. Treatment
66. Symmetrical
67. B.I.B.D.
68. B.I.B.D.
69. Connected, Elementary, Disconnected
70. Design
71. Balanced, Elementary, Precision
72. Balanced connected
73. Unity
74. Efficiency Less
75. Connected
76. Balanced

11.14.3 Multiple Choice Questions and Answers

1.	d	2.	a	3.	c
4.	a	5.	a	6.	b
7.	b	8.	c	9.	c
10.	a	11.	a	12.	c
13.	d	14.	b	15.	d
16.	b	17.	b	18.	b
19.	c	20.	b	21.	c
22.	b	23.	d	24.	d
25.	b	26.	c	27.	a
28.	b	29.	d	30.	a
31.	c	32.	d	33.	b
34.	d	35.	c	36.	c
37.	b	38.	b	39.	c
40.	d	41.	c	42.	d
43.	d	44.	c	45.	b
46.	b	47.	c	48.	c
49.	a	50.	b	51.	b
52.	d	53.	c	54.	b
55.	d	56.	c	57.	d
58.	a				

12

Factorial Experiment

12.1 True / False

Q.1 Factorial experiment is not a design.

Q.2 In factorial experiment, all possible combinations of two or more than two factors are not considered.

Q.3 The different combinations of factors are not treated as different treatments in factorial experiments.

Q.4 Factorial experiment may be conducted in any of the designs say, C.R.D., R.C.B.D., L.S.D. etc.

Q.5 An experiment in which the effect of only one factor say variety or manure is considered at a time, is not known as simple experiment.

Q.6 A factorial experiment in which if (p & q) levels of two factors say (A & B) respectively, are considered together, is called pxq factorial experiment.

Q.7 If the effect of n factors each at two levels is considered in a factorial experiment, it is called 2^n factorial experiment.

Q.8 If two factors A and B are considered in a factorial experiment each at two levels then 1, a, b and ab are considered as different treatment combinations.

Q.9 The mean treatment yields for treatment combination a, b, & ab in a factorial experiment is denoted by (a), (b) & (ab) respectively.

Q.10 Two levels of two factors say A, B in a factorial experiment are denoted by ao, a_1 and bo, b_1 respectively.

Q.11 The two simple effects of A at given two levels bo, & b_1 factors B are deputed respectively as a_1 bo - ao bo and a_1 b_1-ao b_1

Q.12 The average of two simple effects of factor A taken on two different levels of other factor B, is known as "Main Effect" of A.

Q.13 The main effect A is equal to = $\frac{1}{2}(a_1 - ao)(bo + b_1)$

Q.14 Main effect of B is equal to $\frac{1}{4}(b_1 - bo)(ao + a_1)$

Q.15 The interaction effect AB of two factors A & B has been defined by F yates as

$$AB = \frac{1}{2}(a_1 - ao)(b_1 - bo)$$

Q.16 All the simple effects, main effects and interaction effects are known as contrasts or comparision.

Q.17 For the two factors A & B being independent, the interaction effect AB will be of order zero.

Q.18 Interaction effect AB does not give an estimate of interdependence of two factors A & B.

Q.19 The constrasts A, B, & C are not orthogonal to each other in 2^3 factorial experiment.

Q.20 In general if there are n factors say A, B, C , C, __________ etc. each at two levels o and 1 then the main effects and interactions may be represented as

$$x = \frac{1}{2^{n-1}}(a_1 \pm a_0)(b_1 \pm b_o)(c_1 \pm c_o)(d_1 \pm d_o) \text{\_\_\_\_\_\_\_\_\_\_}$$

where the sign in each bracket is negative or positive according if the corresponding capital letter is present or absent in X.

Q.21 If one has to estimate only main effects then both factorial experiment as well as simple experiment give equally efficient estimate.

Q.22 If simple effects of A over various levels of B are declared non significant then there will be interdependence and presence of interactions A B between them.

Q.23 Let there be two factors A & B each at two levels 0 and 1. The treatment's mean yields (i.e. values) are denoted by (1), (a), (b) & (a b)

Q.24 Treatment's totals are denoted by

[1], [a] [b] & [ab]

Q.25 Factorial effect totals are obtained by the expressions.

[A] = [a b] + [a] - [b] - [1]

[B] = [a b] - [a] + [b] - [1]

[AB] = [a b] - [a] - [b] + [1]

12.2 Factorial Experience

Q.26 If there are r replications in 2^2 factorial experiment then SS due to main effect A = $[A]^2 / 2^2 r$ with 1 d.f.

Q.27 SS due to interaction AB = [AB] / 8r with 1 d. f

Q.28 In 2^2 factorial experiment with r replications-

Total d.f = (4r -1)

Error d.f = 3 (r - 1)

Q.29 If F_A, F_B and F_{AB} denote the variance ratio statistics then

$F_B \sim F_{(1, 3(r-1))}$, & $F_{AB} \sim F_{(1, 3(r-1))}$

Q.30 In term of factorial effect total in 2^2 with r replication we have-

Main effect of A= [A]/ 2r

Q.31 In Q(30), the interaction effect AB will be equal to -

Main effected AB = [AB]/2r

Q.32 In Q.30 we have the expression-

S.E ([A] or [B] or [AB]) = $\sqrt{4r\sigma_e^2}$

Q.33 In Q30, we have the expression

$$\text{S.E. (Factorial effect mean)} = \left[\frac{\text{Factorial effect total}}{2r}\right]$$

$$= \sqrt{\frac{4r\sigma_e^2}{4r^2}} = \sqrt{\frac{\sigma_e^2}{r}}$$

Q.34 Let d_1 & d_2 the significant values for factorial effect total and factorial effect mean in 2^2 experiment with r replications then

$$d_1 = t_{\alpha\%, 3(r-1)} \sqrt{4rs^2e}$$

$$d_2 = t_{\alpha\%, 3(r-1)} \sqrt{\frac{-s^2e}{r}}$$

Q.35 F Yates method is opted for not to avoid the specific algebric formule in computation of factorial effect total.

Q.36 The second order interaction ABC is not the difference in interactions AB calculated at each of the two levels of C.

12.3 Experiment in r Replication

Let A, B & C be three factors each at levels 2 then there will $2^3 = 8$ treatment combinations that are 1, a, b, ab, c, ac, bc, abc

Q.37 Let M denotes the mean yield of 8 treatments, then

$$M = \frac{1}{8}\left[(a+1)(b+1)(c+1)\right]$$

Q.38 In 2^3 experiment there will be $(2^3 - 1) = 7$ mutually orthogonal contrasts.

Q.39 In 2^3 experiment with r replication we have Error d.f $= 7(r - 1) = (2^3 - 1)(r - 1)$

Q.40 In 2^3 experiment with r replication we have

S.E (Any factorial effect total) = $\sqrt{2^3 r\sigma_e^2}$

Q.41 In 2^3 experiment with r replications, we have

S.E. (Any factorial effect mean) = $\sqrt{\frac{2^3 r\sigma_e^2}{(4r)^2}} = \sqrt{\frac{\sigma_e^2}{2r}}$

Q.42 In 2^3 experiment with r replication, the interaction effect ABC in term of factorial effect total is
interaction ABC = [ABC]/4r

Q.43 In Question 42 the main effect B, and interaction effect BC in terms of factorial effect total is
Main effect of B = [B]/4r
Interaction effect BC = [BC]/4r

Q.44 In Question 42 the SS due to interaction effect ABC is = $[ABC]^2/2^3r$

Q.45 In 2^n factorial experiment with r replication, we have S. S. due to each factorial effect = $[X]^2/2^n r$ and each will be with 1.d. f and [X] denotes the factorial effect total.

Q.46 In 2^n foactorial experiment with r replication, we have
Error d. f = $(2^n - 1)(r - 1)$

12.4 Confounding Technique

Q.47 Confounding is a device of increasing the block size by making important one or more interaction contrasts identical with block contrast.

Q.48 The design for confounded factorial experiment may be called as "Incomplete Randomised Block Design".

Q.49 Confounding is a device of reducing the block size by making important tone or more interaction contrasts indentical to block contrast.

Q.50 Confounding technique reduces the error variance in a factorial experiment.

Q.50* Confounding technique increases the precision of unconfounded treatment combinations.

Q.51 In a partial confounding, the same interaction is confounded in all the replicates.

Q.52 In a complete confounding, different interactions are confounded in different replicates

Q.53 By sacrificing an information on complete confounded interaction, the unconfounded treatment combination are estimated with greater precision.

Q.54 We can get more information on complete confounded enteraction.

Q.55 If different interactions are confounded in different replicates then such confounding is known as complete confounding.

Q.56 An advantage of partial confounding is that the interaction confounded in one replicate can be estimated from other replicate and vice-versa.

Q.57 In complete confounding, the confounded interaction can always be estimated with probability one.

Q.58 If the number of levels of different factors considered in factorial experiment is not the same then such experiment is known as "Assymmetrical or Mixed Factorial Experiment"

Q.59 The chief advantage of the mixed factorial experiment is that there is more flexibility in considering different levels of different factors according to needs and convenience of experiment.

Q.60 The demerits of a mixed factorial experiment is that the designing and analysis of such experiment are some what complicated.

Q.61 Consider the complete confounding in 2^3 factorial experiment with r replicates, Then

Block d. f = (2r-1)

Q.62 In Question 61, we have
Error d. f = 6(r - 1)

Q.63 In Question 61, for treatment combination, we have
Treatment d.f = (3 - 1) = 6

Q.64 Consider the partial confounding in 2^3 with r replicates, then
Block d.f = (2r - 1)
Treatment d. f = $(2^3 - 1) = 7$

Q.65 In Question 64, we have
Error d.f = 6 (r - 1) -1 and total d.f = 8r - 1

Q.66 Let there be 2^n factorial experiment with block size 2^k. Then number of independent interactions confounded is equal to (n - k)

Q.67 All possible products of (n - k) independent interactions confounded gives some interactions which are known as generalized interactions.

Q.68 The S. S due to all orthogonal contrasts in 2^3 factorial experiment with r replication is equal to. $[\ \ ]^2/2^3r$

Q.69 In case of 2^3 factorial experiment with blocks of size 4, the number of effects can at most be confounded is equal to one

Q.70 The failure of a factor to give the same response at different level of the other factor indicates the presence of interaction.

Q.71 If s^2e is the estimate of error variance σ_e^2 based on n_e d.f then information generated from the experiment is equal to $\frac{s_e^2}{n_e}$

Q.72 If the interactions AB and CD are confounded with blocks in a 2^4 factorial experiment, then automatically confound effect is ABCD

Q.73 The measure of change in the response of a factor A due to change of levels of other factor averaged over all levels of other factor is called Main effect of A.

Q.74 Confounding technique is applicable in case of factorial experiments.

Q.75 A factorial experiment consists of two factors say A and B, at 2 and 3 levels respectively is represented as 2x3 factorial experiment.

Q.76 The idea of factorial experiment was originated by Finney.

Q.77 A factorial experiment with equal number of levels for all factors is known as "Symmetrical Factorial Experiment."

Q.78 Preferably higher order interactions are not selected for confounding.

Q.79 The method of confounding to reduce block size is applicable only for large number of treatment combination in factorial experiments.

Q.80 In 2^4 factorial experiment, the interactions ABC and B.C.D. are confounded to obtain block of size 4 units, the interaction automatically confounded with block is - A.D.

Q.81 If k effects are confounded in a 2^n factorial to have 2^k blocks of size 2^{n-k} units, the number of automatically confounded effect is = $(2^k - k - 1)$.

Q.82 In a 3^3 factorial experiment with three factors A, B & C, each at 3 levels, the interaction A^2BC^2 is same as the interaction $(A^2BC^2)(ABC) = AB^2C$.

12.5 Fill in the Blanks

Q.1 Factorial _________ is not a design.

Q.2 In _________, all the possible combinations of two or more then two factor are considered together.

Q.3 The different _________ of factors are treated as different treatments in _________ experiment.

Q.4 Factorial experiment may be considered in any of the _________ like C. R. D, R. C. B. D and L. S. D

Q.5 In experiment in which the effect of only one factor is considered at a time, is known as _________ experiment.

Q.6 A factorial experiment in which two factors (A & B) are considered at (p and q) levels respectively, is known as _________ experiment.

Q.7 If n factors each at two levels is considered in a factorial _________, it is known as _________ experiment.

Q.8 If n factors each at two levels is considered in a factorial experiment, is known as _________ factorial experiment.

Q.9 In a 2^2 factorial experiment with two factors (A and B), a b, and ab are considered as different treatment _________

Q.10 The mean yield of treatment combinations, a, b, and ab in a factorial experiment are denoted by _________ and (ab) respectively.

Q.11 Two levels of two factors A and B in a factorial experiment is denoted by ao, _________ and , _________, b1 respectively.

Q.12 Two _________ effects of A at given two levels bo and b1 of factor B are denoted by (a_1 bo - ao bo) and _________ respectively.

Q.13 The average of two _________ of factor A taken on two different levels of other B is known as " _________ " of A

Q.14 The main effect of factor $A = \frac{1}{2}($ $)$ (bo _ b1)

Q.15 The main effect of factor $A = \frac{1}{2}($ $)($ $)$.

Q.16 The interaction effect AB is defined as-

Interaction $= \frac{1}{2}$ (_________) (_________)

Q.17 All the simple effects, main effects and interaction effects are known as contrasts or _________

Q.18 For two factors A and B being independent, the interaction effect AB will be of order _________

Q.19 Interaction effect AB gives an estimate of _________ of two factors A and B.

Q.20 The contrasts A, B, & AB are _________ to each other.

Q.21 In general for n factors say A, B, C _________, each at two levels o and 1, the main effects and _________ effects may be represented as-

$$\times = \frac{1}{2^{n-1}}(a_1 \pm a_o)(b_1 \pm b_o)(c_1 \pm c_o) \text{\_\_\_\_\_\_\_\_\_} (\quad)(\quad).$$

Q.22 If simple effects of A over various levels of B are declared non-significants then _________ AB between A and B will be absent.

12.6 2^2 Factorial Experiment

Q.23 Let A and B be two factors each at two levels o and 1. The mean yields (value) are denoted by-

(1), (....), (b) and (......)

Q.24 The treatment total are denoted by-

[1], [ ], [b], and [..............]

Q.25 [B] = [ab] - [.....] + [.......] - [1]
[AB] = [......] - [a] - [.....] + [1]

Q.26 In 2^2 experiment with replications, we have
S. S, due to main effect A = _________ with 1. df

Q.27 S. S (Interaction AB) = _________ with 1. d.f

Q.28 In 2^2 experiment with r replications we have
Total d.f = (_________)
& Error d.f = 3 (_________)

Q.29 Let F_A, F_B and F_{AB} denote the variance ratio statistics then.
$F_A \sim F$ (_________) (_________ and $F_{AB} \sim F_1$(_________), 3 (r - 1)

Q.30 In term of factorial effect total in 2^2 experiment with r replications, we have
Main effect of A= [______] /[______]

Q.31 In Q.30 we have .
Main effect of AB = [______] /[______]

Q.32 In Q.30, we have the expression-
S. E. ([A] or [B] or [AB]) = $\sqrt{(........)}$

Q.33 In Q. 30 we have the expression-
S. E. (Factorial effect mean) = $\sqrt{(........)}$

Q.34 Let d_1 and d_2 denote significant values for factorial effect total and factorial mean effect in 2^2 experiment with r replications then we have
$d_1 = t\alpha\%, 3(.......)\sqrt{(........)}$ and $d_2 = t\alpha\%, 3(.......)\sqrt{(........)}$

Q35. F Yate's method is adopted to avoid to use the specific algebraic formula in computation of _________ effect total.

Q.36 The second order interaction ABC is the difference interaction AB taken at each of the two _________ of _________

12.7 2^3 Experiment in r Replication

Q.37 Let M denotes the mean yield of 8 treatment in 2^3 experiment with three factors A, B, & C with r replications then

$$M = \frac{1}{8}[(a+1)(........)(........)]$$

Q.38 In 2^3 experiment the number of mutually orthogonal contrast is equal to __________

Q.39 In 2^3 experiment with r replications, we have
Error. d.f = _________

Q.40 In 2^3 experiment with r replication, we have

$$\text{S. E. (Any factorial effect mean)} = \sqrt{\frac{\sigma_e^{\ 2}}{(......)}}$$

Q.41 In Q.40 we have

$$\text{S. E. (Any factorial effect mean)} = \sqrt{8..........}$$

Q.42 In Q.40 , in terms of factorial effect total, we have
Interaction ABC = _________ /(4r)

Q.43 In Q.40, in terms of factorial effect total, we have
Main effect of B = [.......]/(4r)
Interaction effect BC = [.......]/(4r)

Q.44 In Q.40, the S. S. due to interaction ABC is equal to = $[ABC]^2$/ (_________)

Q.45 In 2^n factorial experiment with r replications, we have.
S. S (of each factorial effect) = [_______ $]^2$ /_______)
With 1 d.f and [ ] denotes the factorial effect total.

Q.46 In Q.45, we have,
Error d.f = (_______)(_______)

12.8 Confounding Technique

Q.47 Confounding is a device of _______ the block size by making unimportant one or more _______ contrasts identical with block _______

Q.48 The design for ___________ factorial experiment may be called as "___________ Randomised Block Design".

Q.49 Confounding is a device of reducing block size by making ___________ one or more interaction ___________ identical with ___________ contrasts.

Q.50 Confounding technique reduces ___________ variance in a ___________ experiment.

Q.50* Confounding technique increases the ___________ of unconfounding ___________ combinations.

Q.51 In complete confounding the ___________ is confounded in all the ___________

Q.52 In partial confounding, the ___________ treatment are confounded in ___________ replications.

Q.53 By sacrificing the information on ___________ confounded interaction, the unconfounded interactions (treatment combinations) are estimated with ___________ precision.

Q.54 If same interaction is confounded in all the replication then such confounding is known as ___________ confounding.

Q.55 If different interactions are confounded in different replications then such confounding is known as ___________ confounding.

Q.56 An advantage of ___________ confounding is that the interaction confounded in one replication can be ___________ from other replicates.

Q.57 In ___________ confounding, the confounded interaction can not be estimated at all.

Q.58 The chief advantage of assymetrical factorial experiment is that there is more flexibility in choosing the different levels of ___________ factors according to needs and convenience of the experiment.

Q.59 The demerit of a mixed factorial experiment is that the ___________ and ___________ of such experiment is some-what complicated.

Q.60 In complete confounded 2^3 factorial experiment with r replicates the-Block d.f = (___________).

Q.61 In Q.60, we have-

Error d,f = ___________

Q.62 In Q.60, we have for treatment combination-

Treatment d.f= ___________

Q.63 For partial confounding in 2^3 factorial experiment with r replicates, we have

Block d. f = ___________

Q.64 In Q.63, we have

Treatment d.f = ___________

Q.65 In Q.63, we have

Error d.f = ___________

Q.66 In 2^n factorial experiment with block size 2^k, the number of independent interactions confounded is equal to ___________

Q.67 All possible products of (n - k) independent interactions confounded give some more interactios which are known as ___________ interactions.

Q.68 In 2^3 factorial experiment with block of size 4, the number of interaction effects at most confounded is equal to ___________

Q.69 The failure of a factor to give the same response at different levels of other factor indicate the presente of ___________ effect.

Q.70 Let s^2e be the estimate of σ_e^2 based on n_e d.f then infomation generated from the experiment is equal to ___________

Q.71 If the interactions AB and CD are confounded with blocks in a 2^4 factorial experiment then automatically confounded interaction is ___________

Q.72 The measure of rechange in response of a factor averaged over all levels of other factors is known as ___________

Q.73 A factorial experiment with two factors A and B at levels 2 and 3 respectively is denoted as ___________ experiment.

Q.74 The idea of factorial experiment was originated by ___________

Q.75 A factorial experiment with equal number of levels for all factors is known as "___________"

Q.76 Preferably higher order interactions are selected to be ___________ in factorial experiment.

Q.77 Confounding technique to __________ block size is applicable in __________ experiment if the number of treatment combinations is __________

Q.78 In 2^4 factorial experiment, if the interactions ABC and BCD are confound to obtain blocks of size 4 units then automatically confounded interaction with blocks is __________

Q.79 In a 2^n factorial experiment if k effects are to be confounded to obtain 2^k block each of size 2^{n-k} units, the number of automatically confounded interaction is equal to __________

Q.80 In a 3^3 factorial experiment with factors A, B, and C, the interaction $A^2 BC^2$ is same as interaction __________

12.9 Multiple Choice Questions and Answers

Q.1 Factorial experiment is-

(a) Not a design (b) A design

(c) A design and Experiment both (d) None

Q.2 In factorial experiment all possible combinations of two or more factors.

(a) Are considered as treatment

(b) Are not considered as treatment

(c) May or may not be considered as treatment

(d) Are treated as block size.

Q.3 Factorial experiment may be conducted in -

(a) C.R.D. (b) R.C.B.D.

(c) L.S.D. (d) Any of all three

Q.4 An experiment in which the effect of only one factor is considered under study is known as -

(a) Factorial experiment

(b) Mixed experiment

(c) Simple experiment

(d) Symmetrical factorial experiment

Q.5 An experiment in which if two factors A & B are considered together at p and q levels respectively is known as-

(a) pxq factorial experiment

(b) Assymmetrical factorial experiment

(c) Mixed factorial experiment
(d) All the three

Q.6 If two factors A and B are considered in a factorial experiment, each at two level, then treatment combinations are denoted as

(a) 1, a, b, ab, (b) (1), (b), (ab)
(c) [1] [a] [b] [ab] (d) None of three

Q.7 If two factors A and B are under study in 2^2 factorial experiment then mean yield of the treatment combinations a, b and ab will be denoted as

(a) [a] [b] [ab] (b) (a) (b) (ab)
(c) a, b, c (d) None of these

Q.8 Two simple effects of A at given two levels bo and b1 of other factor B will be calculated respectively as

(a) $a_1 b_o - a_o b_o$ and $a_1 b_1$ --- $a_o b_1$
(b) $a_1 b_o - a_1 b_1$ and $a_1 b_1$ --- $a_o b_o$
(c) $a_1 b_1 - a_o b_o$ and $a_o b_o$ --- $a_1 b_1$
(d) $a_1 b_1 - a_o b_o$ and $a_1 b_1$ --- $a_1 b_o$

Q.9 The average of two simple effects of A at two levels bo and b1 of factor B, is known as

(a) Main effect of B (b) Main effect of A
(c) Average effect of A (d) Average effect of B

Q.10 The main effect of A is equal to -

(a) $\frac{1}{2}(a_1 + a_o)(b_1 + b_o)$ (b) $\frac{1}{2}(a_o + a_1)(b_1 - b_o)$

(c) $\frac{1}{2}(a_1 - a_o)(b_1 + b_o)$ (d) $\frac{1}{2}(a_1 - a_o)(b_1 - b_o)$

Q.11 The interaction effect AB in 2^2 factorial experiment is definded as(a)

(a) $AB = \frac{1}{2}(a_o - a_1)(b_o + b_1)$

(b) $AB = \frac{1}{2}(a_1 - a_o)(b_1 - b_o)$

(c) $AB = (a_1 - a_o)(b_1 - b_o)$

(d) $AB = \frac{1}{4}(a_1 - a_o)(b_1 - b_o)$

Q.12 In factorial experiment, all the simple effects, main effects and interation effects are known as-

(a) Contrasts (b) Treatment combination

(c) Comparisons (d) Both (a) and (c)

Q.13 For two factors A and B being independent, the interaction AB will be of order.

(a) One (b) Zero

(c) Two (d) None of these

Q.14 In 2^2 factorial experiment with two factors A and B, the effect AB gives an estimate of-

(a) Interaction effect of A and B

(b) Inter dependence of A and B

(c) Both (a) and (b)

(d) None of these

Q.15 In 2^3 factorial experiment, the contrasts A, B, and C are

(a) Non orthogonal to each other,

(b) Orthogonal to each other

(c) Linearly related to each other

(d) None

Q.16 If simple effects of A over various levels of B in a factorial experiment are declared non-significant then-

(a) Interaction AB will be present

(b) Interaction AB will not be present

(c) Interaction between AB will not be present

(d) Both (a) and (c)

12.9.1: 2^2 Factorial Experiment

Q.17 In 2^2 experiment with treatment A and B, the factorial effect total [B] can be obtained from the expressioin

(a) [B] = [a b] + [a] - [b] - [1] (b) [B] = [a b] - [a] + [b] - [1]

(c) [B] = [a b] + [a] + [b] - [1] (d) [B] = [a b] + [a] - [b] + [1]

Q.18 In Q.17, the factorial effect total for interaction AB is equal to-

(a) [AB] = [a b] - [a] - [b] +[1]
(b) [AB] = [a b] + [a] + [b] - [1]
(c) [AB] = [a b] - [a] + [b] - [1]
(d) [AB] = [a b] - [a] + [b] + [1]

Q.19 In Q.17 with r replication, S. S. due to interaction AB is equal to -

(a) $[AB]^2 / (8r)$
(b) $[AB]^2/(2^2r)$
(c) $[AB]^2 (4r)$
(d) Both (b) and (c)

Q.20 In Q.17 with r replications, error d.f is equal to-

(a) 4 (r - 1)
(b) 3 (r - 1)
(c) 3r - 1
(d) 4r - 1

Q.21 If F_{AB} denotes the variance for interaction AB, in 2^2 experiment then-

(a) $F_{AB} \sim F(1,\ 4(r-1))$
(b) $F_{AB} \sim F(1,\ 3(r-1))$
(c) $F_{AB} \sim F(1,\ (4r-1))$
(d) $F_{AB} \sim F(1,\ 4r-1))$

Q.22 In 2^2 experiment with r replication, in terms of factorial effect total, we have

(a) Main effect of AB - [AB]/ r
(b) Main effect of AB - [AB]/ (2r)
(c) Main effect of AB - $[AB]/(2^2r)$
(d) Main effect of AB - $[AB]^2/(2r)$

Q.23 In Q.17, we have expression for S.E of [A], or [B] or [AB] is equal to

(a) $\sqrt{4r\sigma e^2}$
(b) $\sqrt{2r\sigma e^2}$
(c) $\sqrt[2\sigma e]{r}$
(d) Both (a) and (c)

Q.24 In Q.17, we have expression for S. E. (Factorial effect means)-

(a) $\sqrt{\frac{\sigma e^2}{r}}$
(b) $\sqrt{\frac{\sigma e}{r^2}}$
(c) $\sqrt[\sigma e]{\frac{1}{r}}$
(d) Both (a) and (c)

Q.25 Let in Q.17 d_1 be the significant value for factorial effect totals, then d_1 will be equal to-

(a) $t\alpha, 3(r-1)\sqrt{4_r s_e^2}$ (b) $t\alpha, 3(r-1)\sqrt{4^r s_e}$

(c) $t\alpha, 3(r-1)^2 \sqrt[se]{r}$ (d) Both (a) and (c)

Q.26 Let d_2 be the significant value for factorial effect means in Q.17 then d_2 will be equal to-

(a) $t\alpha, 3(r-1)\sqrt{\frac{s_e^2}{r}}$ (c) $t\alpha, 3(r-1)\frac{Se}{\sqrt{r}}$

(b) Both (a) and (c) (d) $t\alpha, 3(r-1)\sqrt{\frac{2se^2}{r}}$

Q.27 F yate's method is adopted to avoid the specific algebraic formula in computation of-

(a) Factorial effect total, (b) Factorial effect means

(c) S. S due to factorial effects (d) All of the three

12.9.2: 2^3 Factorial Experiment in r Replication

Q.28 Let M be the mean yield of 8 treatment combinations in 2^3 experiment then

(a) $M = \frac{1}{8}[(a+1)(b+1)(c+1)]$

(b) $M = \frac{1}{8}[(a-1)(b-1)(c-1)]$

(c) $M = \frac{1}{8}[(a+b+c+1)$

(d) $M = \frac{1}{8}[(a+b+c)^2]$

Q.29 In 2^3 experiment the number of mutually orthogonal contrasts is equal to-

(a) 8 (b) 7

(c) 6 (d) None

Q.30 In 2^3 experiment with r replication, the error d.f. will be equal to

(a) 6(r - 1) (b) 7 (r - 1)

(c) (8r - 1) (d) (7r - 1)

Q.31 In Question 30, the S.E. (Any factorial effect mean) will be equal to

(a) $\sqrt{\frac{\sigma^2 e}{2r}}$
(b) $\sqrt{\frac{\sigma_e}{2r}}$
(c) $\frac{\sigma_e}{\sqrt{2r}}$
(d) Both (a) & (c)

Q.32 In Question 30, S.E. (Any factorial effect total) will be equal to

(a) $\sqrt{8r\sigma^2 e}$
(b) $\sqrt[2\sigma e]{2r}$
(c) $\sqrt[\sigma e]{8r}$
(d) All the three

Q.33 In Question 30, the interaction effect ABC in term of factorial effect total is equal to

(a) [ABC]/8r
(b) [ABC]/4r
(c) $[ABC]/2^2r$
(d) Both (a) and (b)

Q.34 In Question 30, S.S. due to interaction effect ABC is equal to

(a) $[ABC]^2/2^3r$
(b) $[ABC]^2/8r$
(c) $[ABC]^2/4r$
(d) Both (a) and (b)

Q.35 In 2^n experiment with r replication, S.S. (Due to each factorial effect) will be equal to

(a) $[\ ]^2/2^n r$
(b) $[\ ]^2/8r$
(c) $[\ ]^2/2nr$
(d) None

Where [ ] denote factorial effect totals

Q.36 In Question 30, the error d.f will be equal to

(a) $(2^n - 1)(r - 1)$
(b) $(2n - 1)(r - 1)$
(d) $2^n(r - 1)$
(d) None

12.9.3 Confounding Technique

Q.37 Due to application of confounding technique in factorial experiment the block size is

(a) Increased
(b) Decreased
(c) Unaffected
(d) None

Q.38 Design for confounded factorial experiment may be called as

(a) Incomplete randomised block design
(b) Complete randomised block design
(c) Mixed randomised block design
(d) None of these

Q.39 Due to applying confounding technique in a factorial experiment the error variance is

(a) Unaffected (b) Increased
(c) Decreased (d) None

Q.40 In a confounded factrorial experiment, the precision of unconfounded treatment combination is

(a) Decreased (b) Increased
(c) Unaffected (d) None

Q.41 In a partially confounded experiment, the treatment confounded in all the replications are

(a) Same (b) Different
(c) Mixed (d) None

Q.42 In a complete confounded experiment, the interaction confounded in different replications are

(a) Different (b) Same
(c) Mixed type (d) None

Q.43 In a complete confounded experiment, the precision of unconfounded treatment is

(a) Decreased (b) Increased
(c) Unaffected (d) None

Q.44 In mixed factorial experiment, the different treatments have

(a) No flexibility in number of levels
(b) Flexibility in number of levels
(c) Different precision
(d) Both (b) and (c)

Q.45 In 2^3 factorial experiment with r replication if complete confounding is adopted then block d.f. is equal to

(a) (2r - 1) (b) 2(2r - 1)
(c) 3(r - 1) (d) (3r - 1)

Q.46 In Question 45 the error d. f. is equal to

(a) (6r - 1) (b) 6(r - 1)
(c) 7(r - 1) (d) (7r - 1)

Q.47 In Question 45 the treatment d. f is equal to

(a) 7 (b) 6
(c) 5 (d) None

Q.48 If partial confounding is adopted in 2^3 experiment with r replications, the block d.f is equal to

(a) 2(r - 1) (b) (2r - 1)
(c) 3(r - 1) (d) (3r - 1)

Q.49 In Question 48, the treatment d.f. is equal to

(a) 6 (b) 5
(c) 7 (d) None

Q.50 In Question 48, the error d.f. is equal to

(a) 6(r - 1) (b) (6r - 1)
(c) 6(r - 1)-1 (d) 7(r - 1)

Q.51 In 2^n experiment with block size 2^k, the number of independent interactions confound is equal to

(a) (n - k) (b) (n - k - 1)
(c) (n - k +1) (d) None

Q.52 The S.S. due to all orthogonal contrasts in 2^3 factorial experiment with r replications are equal to

(a) $[\quad]^2/4r$ (b) $[\quad]^2/8r$
(c) $[\quad]^2/2^3r$ (d) Both (b) and (c)

where [ ] denotes the factorial effect total

Q.53 In 2^3 experiment with blocks of size 4, the number of treatment can at most be confounded is equal to -

(a) One (b) Two

(c) Three (d) Four

Q.54 If the interactions AB and CD are confounded with blocks in a 2^4 factorial experiment, then automatically confounded effect is -

(a) ABCD (b) ACD

(d) ADB (d) ABC

Q.55 The measure of change in response of a factor A due to change of levels of other factor averaged over all levels of other factor is known as-

(a) Interaction effect of A with B (b) Main effect of A

(c) Simple effect of A (d) Factorial effect total of A

Q.56 An idea of factorial experiment was originated by-

(a) Finny (b) Fisher R. A

(c) Kempthorm (d) Cox

Q.57 In 2^4 factorial experiment, the interactions ABC and BCD are confounded to obtain block of size 4 units, the interaction automatically confounded is-

(a) AD (b) ABD

(c) BC (d) ACD

Q.58 If k effects are confounded in a 2^n experiment to have 2^k blocks of size 2^{n-k} unit, the number of automatic confoundded effects is equal to-

(a) $(2^k + 1 + K)$ (b) $(2^k - k - 1)$

(b) $(2^k - k + 1)$ (d) $(2^k + k - 1)$

Q.59 In a 3^3 experiment, with three factors A, B, and C, each at three levels the interactions $A^2 B C^2$ is same as-

(a) $AB^2 C$ (b) $A^2 B^2 C$

(c) $B^2 C^2 A$ (d) $A B C$

Q.60 In a factorial experiment, the complete confounded interaction-

(a) Can be estimated with more precision

(b) Can be estimated with less precision

(c) Can not be estimated at all

(d) Nothing can be said with surity

12.10 Key Answers

12.10.1 True / False

1.	True	2.	False	3.	False
4.	True	5.	False	6.	True
7.	True	8.	True	9.	True
10.	True	11.	True	12.	True
13.	True	14.	False	15.	True
16.	True	17.	True	18.	False
19.	False	20.	True	21.	True
22.	False	23.	True	24.	True
25.	True	26.	True	27.	False
28.	True	29.	True	30.	True
31.	True	32.	True	33.	True
34.	True	35.	False	36.	False
37.	True	38.	True	39.	True
40.	True	41.	True	42.	True
43.	True	44.	True	45.	True
46.	True	47.	True	48.	True
49.	False	50.	True	50*.	Ture
51.	False	52.	False	53.	True
54.	False	55.	False	56.	True
57.	False	58.	True	59.	True
60.	True	61.	True	62.	True
63.	True	64.	True	65.	True
66.	True	67.	True	68.	True
69.	True	70.	True	71.	True
72.	True	73.	True	74.	True
75.	True	76.	True	77.	True
78.	False	79.	True	80.	True
81.	True	82.	True		

12.10.2 Fill in the Blanks

1. Experiment
2. Factorial Experiment
3. Combinations Factorial
4. Designs
5. Simple
6. pxq factorial experiment
7. Factorial, 2^n factorial
8. Symmetrical
9. Combinations
10. (a), (b)
11. a_1, b_o
12. Simple $a_1 b_1$ -ao b_1
13. Simple, Main effect
14. (a_1 - ao)
15. (a_1 + ao) (b_1 - bo)
16. (a_1 - ao) (b_1 - bo)
17. Comparisons
18. Zero
19. Interdependence
20. Orthogonal
21. (±) (±)
22. Interaction
23. (a) (ab)
24. a, ab
25. (a) (b)
(ab) (b)
26. $[A]^2/\ 2^2r$
27. $(AB)^2/\ 2^2r$
28. (4r -1), (r - 1)
29. (1), 3(r - 1)

30. [A]/ (2r)
31. [AB] /(2r)
32. 4r σ_e^2
33. (σ_e^2 /r)
34. (r -1), (4r σe^2)

    (r -1), $\left(\frac{s_e^2}{r}\right)$
35. Factorial
36. Levels, C
37. (b+1) (c+1)
38. 7
39. 7(r - 1)
40. (2r)
41. rσ_e^2
42. [ABC]
43. B, BC
44. (2^3 r)
45. (2^n r)
46. (2^n - 1) (r -1)
47. Reducing, Treatment, Contrast
48. Confounded, Incomplete
49. Unimportent Contrast, Block
50. Error , Factorial

50* Precision, Treatment

51. Same Interaction, replications
52. Different, Different
53. Complete, Greater
54. Complete
55. Partial

56. Partial, Estimated
57. Complete
58. Different
59. Design, Analysis
60. (2r -1)
61. 6(r - 1)
62. 6 (six)
63. (2r - 1)
64. $7 = (2^3 - 1)$
65. 6(r - 1)-1
66. (n - k)
67. Generalized
68. One
69. Interaction
70. $\frac{s_e^2}{n_e}$
71. ABCD
72. Main Effect
73. 2x3 Factorial Experiment
74. Finny D. J
75. Symmetrical Factorial
76. Confounded
77. Reduce, Factorial, Large
78. AD
79. $(2^k - k - 1)$
80. $A^5\ BC^5$

12.10.3 Multiple Choice Questions and Answers

1.	a	2.	a	3.	d
4.	c	5.	d	6.	a
7.	b	8.	a	9.	b
10.	c	11.	b	12.	d

13. b	14. c	15. b
16. d	17. b	18. a
19. d	20. b	21. b
22. b	23. d	24. d
25. d	26. c	27. d
28. a	29. b	30. b
31. d	32. d	33. d
34. d	35. a	36. a
37. b	38. a	39. c
40. b	41. b	42. b
43. b	44. d	45. a
46. b	47. a	48. b
49. c	50. c	51. a
52. d	53. a	54. a
55. b	56. a	57. a
58. b	59. a	60. c

13

Split Plot Design

13.1 True / False

Q.1 If two type ot factors, say (A & B) have to be studied in the same experiment where A requires larger plot size and B requires smaller plot size and structure of the experiment is that of plots within blocks and blocks within replication then such layout is known as split plot design.

Q.2 The plots within blocks & blocks within replications are known as "split plot and main plots" respectively.

Q.3 Treatment applied to main plots and split plots are respectively known as split plot treatment and main plot treatment.

Q.4 In split plot experiment, the main plot treatment and split plot treatment are alloted at random to the main plots and split plots in separate way.

Q.5 In split plot experiment, randomisation is done separately for main plot and split plots while in ordinary two factors experiment (factorial), all the treatment combination of two factors are alloted at random to the plots of a block.

Q.6 In split plot experiment, the interaction of main plot treatment and split plot treatment are estimated with more precision than that of main plot treatment but in ordinary two factors experiment 2^2 in R.B.D., all treatments (main effect and interaction) are tested with equal precision.

Q.7 There is similarity of split plot experiment with a confounded factorial experiment.

Q.8 In split plot experiment, the subplots and split plots have the same meaning.

Q.9 In split plot experiment, the terms whole plot treatments and main plot treatment are the two different things.

Q.10 If subplots are considered as plot and whole plots as blocks then the difference among the whole plots are the same as the differences among the levels of the whole plot treatment. Hence we can say that in split plot experiment, the main plot treatments are confound with blocks.

Q.11 Let there be a main factor A at p levels and sub-factor B at q levels with r replications in a split plot experiment. Then statistical model is written as

$$y_{ij} = \mu + r_i + a_j + n_{ij} + b_k + (ab)_{jk} + e_{ijk}$$

$$\text{Where} \begin{pmatrix} i = 1, 2, \ldots\ldots, r \\ j = 1, 2, \ldots\ldots, p \\ k = 1, 2, \ldots\ldots, q \end{pmatrix}$$

n_{ij} ; Main plot error

e_{ijk} : Subplot error

Q.12 The main plot error n_{ij} and subplot error $e_{\lambda jk}$ are distributed independently as

$n_{ij} \sim N(o, \sigma_a^2)$ and $e_{ijk} \sim N(o, \sigma_b^2)$

Q.13 The joint effect (interaction) of split plot treatment and replication is assumed to be zero.

Q.14 There are two types of error in split plot experiment which are known as main plot error (Ea) and subplot Eb.

Q.15 In split plot experiment error "Ea" is smaller than error "Eb).

Q.16 In split plot experiment, subplot factor B is more important

Q.17 Usually the sub factor B an interaction AB are estimated and tested less precisellly than of main factor A

Q.18 The error d.f for the main plot error Ea is equal to (r - 1) (p - 1)

Q.18* An split plot experiment, the d.f for the interaction AB is equal to (p - 1) (q - 1)

Q.19 The error d.f. for subplot treatment B is equal to p(r - 1) (q - 1)

Q.20 The null hypothesis Ho for the main plot treatment is written as

Ho : There is no significant difference between the main plot treatment averaged over all the split plot treatments.

Q.21 The F statistic for main plot treatment A, is defined as

$$F_A = \frac{M.S.A}{Ea} \sim F_{(p-1),(r-1)(p-1)}$$

Q.22 The null hypothesis Ho for the split plot treatment B is written as

Ho : There is no significant difference between subplot treatment averaged over all the main plot treatment.

Q.23 The F statistic to test the significance of split plot treatment B, is defined as

$$FB = \frac{M.S.B}{Ea} \sim F_{(p-1),(r-1)(p-1)}$$

Q.24 The null hypothesis Ho to test the presence of interaction between main plot treatment as subplot treatments is written as-

Ho : The interaction effects between main plot treatment and subplot treatment i. e AB, are all zero.

Q.25 The F statistic to test the presence of interaction AB is defined as-

$$F_{AB} = \frac{M.S\ (AB)}{F_b} \sim F_{(p-1),\ (q-1),\ p(r-1)(q-1)}$$

Q.26 In split plot experiment, the subfactor B rquires smaller experimental material along with main factor A at little extras cost.

Q.27 The $\widehat{S.E}$ (Between two A means e.f (aj - aj')) is equal to $= \sqrt{\frac{2\,Ea}{r\,p}}$

Q.28 The $\widehat{S.E}$ (Between two B means e.f $(b_k - b_{k1})$) $= \sqrt{\frac{2\,Eb}{r\,q}}$

Q.29 The (Between two B means at the same level of A)
e.r different level fo B e.g $(a_1\ b_1 - ao\ b_1)$ or $(a_1\ bo - ao\ bo)$

$$= \sqrt{2\left[\frac{(q-1)\,Eb + Ea}{r.q}\right]}$$

Q.30 If the spliting of the subplot is further continued then the ultimate stage subplot effects would be estimated and tested with the greatest precision.

13.2 Fill in the Blanks

Q.1 In split plot experiment, the main factor A and sub factor B are studied in the ___________ experiment.

Q.2 The factor A require ___________ plot size and factor B requires ___________ plot size in ___________ experiment.

Q.3 The structure of split plot experiment is that of the split plots ___________ blocks and blocks ___________ replicates.

Q.4 The plots within blocks and blocks within replicates are known as ___________ and ___________ respectiely.

Q.5 Treatments applied to ___________ and ___________ are known respectively as main plot treatment and subplot treat ment.

Q.6 In split plot experiment, the main plot treatments and split plot treatment are applied at ___________ to the ___________ and ___________ independentally.

Q.7 In factorial experiment, all the treatment combinations of two or more than ___________ factors are alloted at ___________ to the plots ___________ blocks.

Q.8 In split plot experiment the interaction AB are estimated with ___________ than the factor A and B respectively.

Q.9 In ordinary 2^2 factorial experiment in R.C.B.D., all treatment (main and interaction) are estimated and ___________ with the ___________ precision.

Q.10 There is similarity in ___________ experiment and in confounded factorial experiment.

Q.11 In split plot experiment the subplot treatment and ___________, both are same.

Q.12 In split experiment, the whole plot treatment and main plot treatment indicate the same treatment.

Q.13 Let there be a main factor A and sub factor B at the levels p & q respectively with r replications, in a split plot experiment, then statistical model is written as-

$$y_{ijk} = \mu + \ldots\ldots + \ldots\ldots + \ldots\ldots + \ldots\ldots + \ldots\ldots + \ldots\ldots$$

$$\text{Where} \begin{pmatrix} i = 1, 2, \ldots\ldots, r \\ j = 1, 2, \ldots\ldots, p \\ k = 1, 2, \ldots\ldots, q \end{pmatrix}$$

Q.14 The main plot error n_{ij} and subplot error eijk are distributed independently as-

$n_{ij} \sim N$ (___________, ___________), eij $\sim N$ (___________, σa^2)

Q.15 In split plot experiments, the interaction (Replications and split plot treatment) effect is assumed to be equal to ....

Q.16 Two type of experimental errors in split plot design say Ea and Eb are respectively known as ___________ and ___________

Q.17 In split plot experiment, error Eb is ___________ than the main plot error Ea.

Q.18 In split plot experiment, the error ___________ is more important than the error ___________

Q.19 Usualy the _______ B and AB are estimated and tested ________ precisely

Q.20 The error d.f of the main plot error Ea is equal to ___________

Q.21 The error d.f for the interaction AB in split plot experiment is equal to ___________

Q.22 In split plot experiment, the d.f for interaction AB is equal to __________

Q.23 The null-hypothesis for main plot treatment is written as "There is no significant difference between main plot treatment ___________ over all the split plot treatments.

Q.24 The F statistic for main plot treatment A is defined as-

$$F_A = \frac{M.S.A}{\dots\dots\dots} \sim F_{(\dots),(\dots)(\dots)}$$

Q.25 The null hypothesis for the treatment B is written as that "There is no ___________ difference between subplot treatments averaged over all ___________

Q.26 The F_B statistic to test the significance of subplot treatment is defined as-

$$F_B = \frac{M.S.B}{\dots\dots\dots} \sim F_{(\dots),-(\dots)(\dots)}$$

Q.27 The null hypothesis for the presence of interaction AB is written as "The interaction effects AB are all equal to ___________"

Q.28 The F_{AB} statistic to test the presence of interaction effect AB is written as-

$$F_{AB} = \frac{M.S(AB)}{..............} \sim F_{(....)(....),-(....)(....)}$$

Q.29 In split experiment, the subfactor B requires __________ experiment material than that of main __________ A at little extra cost.

Q.30 The estimate of S. E. (between two A means) is equal to = __________

Q.31 The estimate of S.E (between two B means) = __________

Q.32 The estimate of S. E (between two B means) at the same level of A or different level of B)

$$= \sqrt{2\frac{[(.....)\ E\,b+......]}{r\,q}}$$

Q.33 If the spliting of the sub factor is further continued then the ultimate subplot effects would be estimated and tested with the __________ precision.

13.3 Multiple Choice Questions and Answers

Q.1 In the split plot experiment, the main factor A and subfactor B are studied in the __________

(a) Factorial experiment (b) Same experiment
(c) Two simple experiment (d) Latin square design.

Q.2 The main factor A requires

(a) Smaller plot size (b) Squared plot size
(c) Larger plot size (d) Sub plot

Q.3 The sub plot treatment B requires

(a) Larger plot size (b) Smaller plot size
(c) Sub plots (d) Both (b) and (c)

Q.4 In split plot experiment, the subfactor B is alloted to the split plots

(a) Within blocks (b) Between blocks
(c) Between rows (d) Between columns

Q.5 In split plot experiment, the factors alloted to plots within blocks are known as

(a) Subfactors (b) Split plot treatment

(c) Interactions (d) Both (a) and (b)

Q.6 In split plot experiment, the factors alloted to blocks within replication, are known as

(a) Main plot treatment (b) Whole plot treatment

(c) Both (a) and (b) (d) Subplot treatment

Q.7 In factorial experiment, all the treatment combinations of two or more than two factors are alloted at random to the

(a) Plots within blocks

(b) Block within replications

(c) Split plots within main

(d) Split plots within replications

Q.8 In split plot experiment, the maximum importance is given to the

(a) Main factor A (b) Sub factor B

(c) Interaction A B (d) Interaction RA

Q.9 In split plot experiment, the precision of interaction AB is

(a) Equal to that of A

(b) Equal to that of B

(c) More than that of A and B both

(d) Less than that of A and B

Q.10 In ordinary 2^2 factorial experiment in R. C.B .D, the interaction effect AB will be estimated with precision

(a) Equal to that of A

(b) Les than that of A

(c) Equal to that of A and B both

(d) More than that of A & B both

Q.11 In split plot experiment, the subplot treatment is

(a) Different from main plot treatment

(b) Different from whole plot treatment

(c) Same as split plot treatment

(d) All the (a), (b) & (c) are correct

Q.12 In split plot experiment if main plot error variance and sub plot error variance are as σ_a^2 and σ_b^2 respectively then the error term n_{ij} for the main factor A is distributed as

(a) $n_{ij} \sim N(o, \sigma_b^2)$ (b) $n_{ij} \sim N(\mu, \sigma_a^2)$

(c) $n_{ij} \sim N(o, \sigma_a^2)$ (d) $nij \sim N(\mu, \sigma_b^2)$

Q.13 In question 12, the error term e_{ijk} occuring in subplot treatment B, is distributed as

(a) $eijk \sim N(\mu, \sigma_b^2)$ (b) $eijk \sim N(0, \sigma_a^2)$

(c) $eijk \sim N(\mu, \sigma_a^2)$ (d) $eijk \sim N(0, \sigma_b^2)$

Q.14 In a split plot experiment, between the main plot error Ea and subplot error Eb, the following relation exists

(a) Ea < Eb, (b) Ea = Eb

(c) Ea > Eb (d) None

Q.15 In split plot experiment, the d.f. for the error Ea is equal to

(a) (r -1) (q - 1) (b) (r - 1) (p - 1)

(c) (p - 1) (q - 1) (d) (p - 1)

Q.16 In split plot experiment the d.f. for interaction AB is equal to

(a) (r -1) (q - 1) (b) (r - 1) (p - 1)

(c) (p - 1) (q - 1) (d) p (q - 1)

Q.17 In split plot experiment the d.f for the error E_b is equal to

(a) p (r -1) (q - 1) (b) r (p - 1) (q - 1)

(c) q (r - 1) (p - 1) (d) None

Q.18 The F_A statistic defined for main plot treatment A is distributed as

(a) $F_A \sim {}^F(q - 1), (r - 1)(p - 1)$

(b) $F_A \sim {}^F(p - 1), (r - 1)(p - 1)$

(c) $F_A \sim {}^F(p - 1), p(r - 1)(p - 1)$

(d) $F_A \sim {}^F(q - 1), p(r - 1)(q - 1)$

Q.19 The F_B statistic defined for sub factor B is distributed as

(a) $F_B \sim {}^F(q - 1), (r - 1)(p - 1)$

(b) $F_B \sim {}^F(q - 1), (p - 1)(q - 1)$

(c) $F_B \sim F\,(q - 1),\ p\,(r - 1)\,(q - 1)$

(d) $F_B \sim F\,(p - 1),\ (r - 1)\,(q - 1)$

Q.20 F_{AB} statistic defined for interaction AB in split plot experiment is distributed as

(a) $F_{AB} \sim F\,(p - 1)\,(q - 1),\ p\,(r - 1)\,(q - 1)$

(b) $F_{AB} \sim F\,(p - 1)\,(r - 1),\ r\,(p - 1)\,(q - 1)$

(c) $F_{AB} \sim F\,(p - 1)\,(q - 1),\ q\,(r - 1)\,(p - 1)$

(d) $F_{AB} \sim F\,(r - 1),\ (q - 1)\,p\,(r - 1)\,(q - 1)$

Q.21 The estimate of S.E (Between two A means) is equal to

(a) $\sqrt{\frac{2E_a}{rp}}$ (b) $\sqrt{\frac{2E_a}{rq}}$

(b) $\sqrt{\frac{2E_a}{pq}}$ (d) $\sqrt{\frac{2E_b}{rq}}$

Q.22 The estimate of S.E. (Between two B means) is equal to

(a) $\sqrt{\frac{2E_b}{rq}}$ (b) $\sqrt{\frac{2E_b}{pq}}$

(c) $\sqrt{\frac{2E_b}{rq}}$ (b) $\sqrt{\frac{2E_a}{rp}}$

Q.23 The estimate of S.E. (Between two B means are the same level of A) is-

(a) $\sqrt{\frac{2E_b}{r}}$ (b) $\sqrt{\frac{2E_a}{r}}$

(c) $\sqrt{\frac{2E_b}{p}}$ (d) $\sqrt{\frac{2E_b}{q}}$

Q.24 S.E. (Difference between two A means at the same level of B or different levels of B) is equal to

(a) $\sqrt{2\left[\frac{(q-1)E_b+E_a}{rq}\right]}$ (b) $\sqrt{2\left[\frac{(q-1)E_b+E_a}{rq}\right]}$

(c) $\sqrt{\left[\frac{(p-1)E_b+E_a}{rp}\right]}$ (d) $\sqrt{\left[\frac{(p-1)E_b+E_a}{rq}\right]}$

13.4 Key Answers

13.4.1 True / False

1.	True	2.	False	3.	False
4.	True	5.	True	6.	True
7.	True	8.	True	9.	False
10.	True	11.	True	12.	False
13.	True	14.	True	15.	False
16.	True	17.	False	18.	True
18*	True	19.	True	20.	True
21.	True	22.	True	23.	True
24.	True	25.	True	26.	True
27.	False	28.	False	29.	True
30.	True				

13.4.2 Fill in the Blanks

1. Same
2. Large, smaller, split plot
3. Within, within
4. Subplots, main plots
5. Main plots, sub-plots
6. Random, main plots, sub-plots
7. Two, Random, Within
8. More precision
9. Tested, same
10. Split plot
11. Split plot treatments
12. Same Type
13. r_i, aj, n_{ij}, bk, (a b)jk, eijk
14. N (O, σ_a^2) (O, σ_b^2)
15. Zero
16. Main plot error sub-plot error
17. Less
18. E.b, Ea

19. Subfactor, Interaction
20. (r - 1) (p - 1)
21. p(r - 1) (q - 1)
22. (p - 1) (q - 1)
23. Averaged
24. Ea , (p - 1), (r -1) (p - 1)
25. Significant, main plot treatment
26. Eb, (q - 1), p (r - 1) (q - 1)
27. Zero
28. Eb, (p - 1) (q - 1), p(r - 1) (q - 1)
29. Smaller, factor
30. $\sqrt{\frac{2E_a}{rq}}$
31. $\sqrt{\frac{2E_b}{rq}}$
32. (q - 1), Ea
33. Greatest

13.4.3. Multiple Choice Questions and Answers

1.	b	2.	c	3.	d
4.	a	5.	d	6.	c
7.	a	8.	c	9.	c
10.	c	11.	d	12.	c
13.	d	14.	c	15.	b
16.	c	17.	a	18.	b
19.	c	20.	a	21.	b
22.	c	23.	a	24.	a

14

Strip Plot Design

14.1 True/False

Q.1 Let there be two factors A and B with p and q levels respectively which are applied to the larger strip by dividing the experimental material into p rows and q columns. The factors A and B are applied at random to these rows and columns respectively. Such a layout is known as strip plot design.

Q.2 If the two factors A and B requirs larger plots and interaction AB is considered more important than the main factors A and B both then strip plot design is suitable.

Q.3 Two set of treatments say (i) fertilizers, tillage operation and (ii) spraying and tillage treatment with ordinary form implements may be treated as two factors A and B to be applied to strip plot design.

Q.4 The split block design, two way whole plot design, two way main plot design etc are the other appropriate terms used for denoting the strip plot design.

Q.5 The statistical model used for strip plot design is written as $y_{ijk} = \mu + r_i + a_j + (ra)_{ij} + b_k + (r\,b)_{ik} + (ab)_{jk} + e_{ijk}$ in usual notations.

Q.6 The assumptions on the effects ri, aj and b_k are as

$$\sum_{i=1}^{r} ri \neq \sum_{j=1}^{p} aj \neq \sum_{k=1}^{q} b_K \neq 0$$

Q.7 Statistical assumptions on interaction effects $(ra)_{ij}$, $(r.b)_{JK}$ in and $(ab)_{Jk}$ are as-

$$\sum_{ij} (ra)_{ij} \neq \sum (r\,b)_{jk} \neq \sum_{jk} (ab)\,jk \neq 0$$

Q.8 There are three experimental error variances denoted as Ea, Eb, and Ec which occur in strip plot design.

Q.9 The experimental error Ea is used to test the significance of main effect B.

Q.10 The experimental error Eb is useful to test the significance of main factor A.

Q.11 The experiment error Ec is used to test the significance of interaction effect AB.

Q.12 The null hypothesis to test the significance of factor A is Ho : aj = 0 ie there is no significant difference between two main effects of factor A averaged over all the q levels of second factor B.

Q.13 The Fa statistic to test the significance of main factor A is defined and follows the distribution as

$$F_A = \frac{M.S.A}{Ea} \sim {}^{F}(p-1),(r-1)(p-1)$$

Q.14 The null hypothesis to test the significance of main effect of factor B is Ho : b_k = 0 ie there is no significant difference between two (main effects) of the factor B averaged over all the p levels of the factor A

Q.15 The F_b statistic to test the significance of main factor B is defined as and follows the distribution as

$$F_B = \frac{M.S.B}{E_b} \sim {}^{F}(q-1),(r-1)(q-1)$$

Q.16 The null hypothesis for the interaction AB is

Ho: (ab)jk = 0, ie the two factors A and B have no interdependence (or interaction effect).

Q.17 The test statistic F_{AB} for the interaction AB is defined and follows the F distributions as-

$$F_{AB} = \frac{MS(AB)}{Ec} : F(p-1)(q-1),(r-1)(p-1)(q-1)$$

Q.18 Estimate of S E. (Difference between main effects of factor A is equal to-

$$\sqrt{\frac{2\,Ea}{rq}}$$

Q.19 $\widehat{S.E}$ (Difference between two main effects of factor B) is

$$\sqrt{\frac{2E_b}{rp}}$$

Q.20 $\widehat{S.E}$ (Difference between two A means at the same level of B)

$$= \sqrt{2\left[\frac{Ea+(q-1)Ec}{rq}\right]}$$

Q.21 $\widehat{S.E}$ (Difference between two B means at the same level of A)

$$= \sqrt{2\left[\frac{Eb+(p-1)Ec}{rp}\right]}$$

Q.22 Out of three error variances Ea, Eb and Ec, the largest error variance is the Ec.

Q.23 Out of three type of effects A, B and AB, the least precisely estimated and tested effect is the "Interactions AB.

Q.24 The d.f for the error variance Ea = (r -1)(q - 1)

Q.25 The d.f for the error variance Eb = (r -1)(p - 1)

Q.26 The d.f for the error variance Ec = (r -1)(p - 1)(q - 1)

Q.27 Out of three type of error d.f corresponding to error variance Ea, Eb and Ec, the largest d.f corresponds to the error variance "Ec".

Q.28 Out of three type of effects A, B and AB, the most precisely estimated and tested effect is "Interaction AB".

14.2 Fill in the Blanks

Q.1 In a strip plot design, two factors A and B with p and q levels respectively, are applied to larger strips prepared into p rows and ____________

Q.2 In strip plot design, the two factors A and B are applied at ____________ to these p ____________ and q ____________ respectively.

Q.3 If two factors A and B require larger plots and interaction AB is to be tested more precisely than A and B both, then suitable design used is ____________

Q.4 The split block design or two way main plot design are other names used for design known as ____________

Q.5 The assumptions on replication effect r_i, main effect aj of factor A and main effect bk of factor B in strip plot design is.

$$\sum_{i=1}^{r} r_i = \sum_{j=}^{p} a_j = \sum_{k=}^{q} \ldots\ldots\ldots = \ldots\ldots\ldots$$

Q.6 The assumptions on interaction effect (ra)ij, (rb)ik and (ab)jk is strip plot design is-

$$\sum_{ij}(ra)_{ij} = ... = \sum_{jk}(ab)_{jk} = ...$$

Q.7 The three error variances in ____________ design are denoted as Ea, ____________ and Ec respectively.

Q.8 The experimental error variance Ea is used to test the significant difference between two main effects of the factor ____________

Q.9 The error variance Eb is applicable to test the significant difference between to main ____________ of the factor ____________

Q.10 The experimental error Ec is applicable to test the ____________ difference between two ____________ effects AB.

Q.11 Ho: aj, There is no significant difference between two ____________ of the factor A _________ over all the q levels of factor __________

Q.12 Ho: bk = 0, is to test the significant difference between two ____________ of the factor ____________ averaged over all the ____________ levels of factor A.

Q.13 Ho: $(ab)_{jk} = 0$, is used to test the ____________ of interdependence between two factors ____________ and ____________

Q.14 The test statistic defined as

$$FA = \frac{M.S.A}{Ea} \sim F(......), (r-1)(......)$$

Q.15 The test statistic F_b is defined as

$$F_b = \frac{M.S.B}{Eb} \sim F(........)\ (r - 1)(........).$$

Q.16 The test statistic F_{AB} to test the significance of interaction AB is defined and follows the distribution as-

$$F_{AB} = \frac{M.S.(A.B)}{Ec} \sim F\ (p - 1)\ (.......),\ (........)\ (p - 1)\ (.......)$$

Q.17 $\widehat{S.E.}$(Difference between two main effects of factor A) is equal to

$$= \sqrt{\frac{2....}{r....}}$$

Q.18 $\widehat{S.E.}$ (Difference between two main effects of factor B) is

$$=\sqrt{\frac{2Eb}{........}}$$

Q.19 $\widehat{S.E.}$ (Difference between two A means at the same level of B) is-

$$=\sqrt{2\left[\frac{Ea+(......)......}{r\,q}\right]}$$

Q.20 $\widehat{S.E.}$ (Difference between two B means at the same level of A) is-

$$=\sqrt{2\left[\frac{......+(p-1)......}{r\,p}\right]}$$

Q.21 Out of three error variances Ea, Eb and Ec the smallest error variance is the ____________

Q.22 Out of three type of effects A, B and AB, the most precisely estimated and tested effect is the ....

Q.23 The error d.f. for the error Ea = (......) (......)

Q.24 The error d.f. for the error Eb = (......) (.......)

Q.25 The error d.f. for the interaction AB = (.......) (.......) (.......)

Q.26 Out of three errors Ea, Eb and Ec, the largest d.f belongs to the error ______________

14.3 Multiple Choice Questions and Answers

Q.1 In a strip plot design two factors A and B with p & q levels respectively are applied to

(a) Larger strips (b) Smaller stirps
(c) Splited strips (d) Sub strips

Q.2 In strip plot design, the two factors A and B are applied to p rows and q columns respectively in a

(a) Systematic way (b) Random way
(c) Purposive way (d) None of these

Q.3 If two factor A and B require larger plots, and interaction AB is to be tested more precisely than A and B both, then the suitable design applicable is the

(a) Split plot design (b) Split block design
(c) Strip plot design (d) Both (a) and (c)

Q.4 The other suitable names for strip plot design may be as

(a) Split block design
(b) Two way main plot design
(c) Two way whole plot design
(d) All the three

Q.5 In strip plot design, the assumption on effects ri, aj and bk of replication, two factor A and B are

(a) $\sum_{i}^{r} r_i = 0$ (b) $\sum_{j=1}^{p} aj \neq 0$

(c) $\sum_{k=1}^{q} bk = 0$ (d) $\sum_{i}^{r} r_i = \sum_{j}^{p} aj = \sum_{k}^{q} bk = 0$

Q.6 The statistical assumptions on interaction effects $(ra)_{ij}$, $(rb)_{ik}$ and $(ab)_{jk}$ in strip plot design as

(a) $\sum_{ij} (ra)_{ij} = 0$

(b) $\sum_{jk} (ab)_{jk} \neq 0$

(c) $\sum_{jk} (rb)_{ik} \neq 0$

(d) $\sum_{ij} (ra)_{ij} = \sum_{ik} (rb)_{ik} = \sum_{jk} (ab)_{jk} = 0$

Q.7 In strip plot design for the factor A, B, and AB, the error variance respectively are denoted as

(a) Ea, Ec Eb, (b) Ea, Ec, Ea, Ed
(c) Ea, Eb, Ec (d) Ec, Eb, Ea

Q.8 The error variance Ea is used to test the significant difference between two

(a) Main effects of factor B (b) Main effcet of factor A
(c) Interaction effects AB (d) None of these

Q.9 The error variance Ec is applicable to test the significant difference between two

(a) Main effect of the factor B (b) Interaction effect AB
(c) Main effects of factor A (d) None of these

Q.10 The error variance Eb is used to test the significant difference between two

(a) Main effect of the factor B (b) Interaction effect AB

(c) Main effects of factor A (d) None of these

Q.11 Ho aj =0 implies that there is no significant differene between two main effect of factor A averaged over all the

(a) q levels of factor A (b) q levels of factor B

(c) p levels of factor A (d) p levels of factor B

Q.12 Ho bk=0 implies that there is no significant difference between two main effect of factor B averaged over all the

(a) p levels of factor A (b) q levels of factor B

(c) p levels of factor B (d) q levels of factor A

Q.13 The test statistic F_A is defined and distrubuted as

(a) $F_A = \dfrac{MSA}{E_a} \sim {}^{F}(p-1),(r-1)(p-1)$

(b) $F_A = \dfrac{MS.A}{E_a} \sim {}^{F}(q-1),(r-1)(p-1)$

(c) $F_A = \dfrac{M.SA}{E_b} \sim {}^{F}(q-1),(r-1)(p-1)$

(d) $F_A = \dfrac{M.SA}{E_b} \sim {}^{F}(p-1),(r-1)(q-1)$

Q.14 The test statistic F_b is defined and distributed as-

(a) $F_b = \dfrac{MSB}{E_b} \sim {}^{F}(p-1),(r-1)(q-1)$

(b) $F_b = \dfrac{MS.B}{E_b} \sim {}^{F}(q-1),(r-1)(q-1)$

(c) $F_b = \dfrac{M.SB}{E_a} \sim {}^{F}(q-1),(r-1)(q-1)$

(d) $F_b = \dfrac{M.SB}{E_B} \sim {}^{F}(p-1),(q-1)(r-1)(p-1)(q-1)$

Q.15 The test statistic FAB for testing interaction AB is defined and distributed as-

(a) $F_{AB} = \dfrac{M.S.(AB)}{E_c} \sim {}^{F}(p-1)(q-1),(r-1)(p-1)(q-1)$

(b) $F_{AB} = \dfrac{M\ S\ (AB)}{E_c} \sim {}^{F}(r-1)(p-1),(r-1)(p-1)(q-1)$

(c) $F_{AB} = \dfrac{M.S\ (AB)}{E_c} \sim {}^{F}(r-1)(q-1),(r-1)(p-1)(q-1)$

(d) $F_{AB} = \dfrac{M.S.\ (AB)}{E_B} \sim {}^{F}(p-1)(p-1),(r-1)(p-1)(q-1)$

Q.16 $\widehat{S.E}$ (Difference between two main effects of A) is equal to -

(a) $\sqrt{\dfrac{2\,E_a}{r\,p}}$ (b) $\sqrt{\dfrac{2\,E_a}{r\,q}}$

(c) $\sqrt{\dfrac{2\,E_b}{r\,p}}$ (d) $\sqrt{\dfrac{2\,E_b}{r\,q}}$

Q.17 $\widehat{S.E}$ (Difference between two main effect B) is equal to-

(a) $\sqrt{\dfrac{2\,E_b}{r\,q}}$ (b) $\sqrt{\dfrac{2\,E_a}{r\,p}}$

(c) $\sqrt{\dfrac{2\,E_b}{r\,p}}$ (d) $\sqrt{\dfrac{2\,E_a}{r\,q}}$

Q.18 $\widehat{S.E}$ (Difference between two A means at the same level of B) is equal to-

(a) $\sqrt{2\left[\dfrac{E_a+(q-1)E_c}{r\,q}\right]}$ (b) $\sqrt{2\left[\dfrac{E_a+(p-1)E_c}{r\,q}\right]}$

(c) $\sqrt{2\left[\dfrac{E_a+(q-1)E_c}{r\,p}\right]}$ (d) $\sqrt{2\left[\dfrac{E_a+(q-1)E_b}{r\,q}\right]}$

Q.19 $\widehat{S.E.}$ ((Difference between two B means at the same level of A) is equal to-

(a) $\sqrt{\frac{2[E_b + (p-1)E_c]}{rp}}$ (b) $\sqrt{\frac{2[E_b + (p-1)E_c]}{rq}}$

(c) $\sqrt{\frac{2[E_a + (p-1)E_c]}{rp}}$ (d) $\sqrt{\frac{2[E_a + (p-1)E_c]}{rq}}$

Q.20 Out of three error variances Ea, Eb and Ec, the smallest error variance is-

(a) Ea (b) Eb

(c) Ec (d) Any one of the three

Q.21 Out of three type of effect A, B, and AB, the most precisely estimated and tested effect is

(a) A (b) B

(c) AB (d) Cannot be decided

Q.22 The error d.f. for the error Ea is equal to

(a) (r - 1) (q - 1) (b) (r -1) (p - 1)

(c) (p -1) (q - 1) (d) (r -1) (p - 1) (q -1)

Q.23 The error d.f for the error Eb is equal to

(a) (r - 1) (q - 1) (b) (r -1) (p - 1)

(c) (p -1) (q - 1) (d) (r -1) (p - 1) (q -1)

Q.24 The error d.f for the interaction AB is equal to

(a) (r - 1) (p - 1) (b) (r -1) (q - 1)

(c) (p -1) (q - 1) (d) (r -1) (p - 1) (q -1)

Q.25 Out of three error Ea, Eb and Ec, the largest d.f corresponds to the error

(a) E_a (b) E_b

(c) E_c (d) Can not be decided

14.4 Key Answers

14.4.1 True / False

1.	True	2.	True	3.	True
4.	True	5.	True	6.	False
7.	False	8.	True	9.	False
10.	False	11.	True	12.	False
13.	True	14.	True	15.	True
16.	True	17.	False	18.	True
19.	True	20.	False	21.	True
22.	False	23.	False	24.	False
25.	False	26.	True	27.	True
28.	True				

14.4.2 Fill in the Blanks

1. q Columns
2. Random, Levels (rows), Levels (columns)
3. Strip plot design
4. Strip plot design
5. b_K, Zero
6. $\sum_{ik}(r\,b)ik$, zero
7. Strip plot design, Eb
8. A
9. Effect, B
10. Significant, interaction
11. Main effect, Averaged, B
12. Main effect , B, p
13. Presence, A, B
14. (p - 1), (p - 1)
15. (q - 1), (q - 1)
16. (q -), (r - 1) (q - 1)

17. Ea / q
18. r p
19. (q - 1), Ec
20. Eb, Ec
21. Ec
22. AB
23. (r - 1) (p - 1)
24. (r - 1) (q - 1)
25. (r - 1) (q - 1) (q - 1)
26. Ec

14.4.3 Multiple Choice Questions and Answers

1.	a	2.	b	3.	d
4.	d	5.	d	6.	d
7.	c	8.	b	9.	b
10.	a	11.	b	12.	a
13.	a	14.	b	15.	a
16.	b	17.	c	18.	a
19.	a	20.	c	21.	c
22.	b	23.	a	24.	d
25.	c				

15

Progeny Row Trial and Compact Family Block Design

15.1 True / False

Q.1 The replicated experiment designed to compare a number of parent plants w. r. to some attribute under study is known as "Progeny row trial?

Q.2 During old days, the mass selection technique was adopted in plant breeding to select the plants showing superior genetic values.

Q.3 Now progeny row trial has been investigated by plant breeders as an improved technique over mass selection technique.

Q.4 In a progeny row trial each plot consists of a single row and all the seeds of the row belong to the same parent plant.

Q.5 A progeny row trial is carried out in a simple R. B. D. in which each plot is a progeny row and the various progenies to be tested are randomised in each block.

Q.6 Let there be p progenies and r replications with ki plants in each plot. Then d. f. for progenies is equal to (r - 1).

Q.7 In Question 6, the error (Plot error) d.f. = (r -1) (p - 1)

Q.8 In Question 6, the d.f for pooled plant error = $\sum_{i}^{rp} (k_i - 1)$

Q.9 Progeny row trial has been found suitable for the crops, like wheat, cotton, sorghum & groundnut.

Q.10 If the progenies tried in an experiment belong to a number of strains or families etc. then two types of comparisons are involved.

(i) The comparison between different families

(ii) Comparision between progenies within family. A layout suitable for such study is known as "Compact family block design".

Q.11 In compact family block design, the progenies belonging to the same family are sown side by side in a family main plot.

Q.12 In compact family block design, the family plots are randomised in each block.

Q.13 The progeny plots are randomised within the main plots (i.e. family plots) and constitute the subplots.

Q.14 Compact family block design is quite analogous to split plot design.

Q.15 Let there be f families, p progenies in each family and r replications in the compact family block design, then- the error d.f of the error Ea for the main plot = (r - 1)(f - 1)

Q.16 In Question 15, the d.f. for progenies within family is = f (p -1)

Q.17 The d.f. for the error Eb i.e for subplot error = (f p r -1)

Q.18 In compact family block design, the subplot treatments are different for all the main plots where as in split plot design, the sub plot treatments are same for all the main plots.

Q.19 S.E (Difference between two progency means taken from two different families)

$$= \sqrt{\frac{2}{r}\left[\frac{Ea + (p+1)\,Eb}{r}\right]}$$

Q.20 To test the homogeneity of error variances obtained from family to family, the test used is the " Bartlett test of homogeneity of variances.

15.2 Fill in the Blank

Q.1 The replicated experiment designed to compare a number of parent plants w. r. to some characteristic under study is known as "______________".

Q.2 The mass selection technique is adopted in plant breeding to select the plants showing _____________ values.

Q.3 Progeny row trial investigated by plant breeder is an improved technique over "_____________ technique".

Q.4 In progeny row trial each plot contains a _____________ and all seeds of the row belong to the same _____________

Q.5 A progeny row trial is conducted in a simple _____________ in which each plot is a _____________ and various progenies under test are randomised in _____________

Q.6 Let there be progenies wtih r replication and ki plants in each plot then d. f. for the progenies is equal to ____________

Q.7 In Question 6. the d.f. for plot error = ____________

Q.8 In Question 6 the pooled plant' error d.f. = ____________

Q.9 If the progenies tested in a experiment belong to a number of strains or families, then ____________ of comparisons are studied.

Q.10 In progeny row trials, the two types of comparisons considered are between different ____________ and between progenies ____________ family.

Q.11 A layout suitable for studies as considered in Question 10, is known as "____________".

Q.12 In a compact family block design, the progenies belonging to the ____________ are sown side by side in a family ____________ plot.

Q.13 In compact family block design, the family plots are ____________ in each block.

Q.14 The progeny plots are randomised ____________ the family plots and constitute the ____________

Q.15 The compact family block design is quite analogous to "____________".

Q.16 Let there be f families, p progenies in each family and r replication in a compact family block design. Then the error (Ea) d.f for the main plot = ____________

Q.17 In Question 16, the d.f. for progenies within family is ____________

Q.18 In Question 16, the d.f. for the error E_b i.e for sub plot error is ________

Q.19 In compact family block design, the sub plot treatments are = ________

Q.20 S.E. (Difference between two progeny means taken from two different family)

$$= \sqrt{\frac{2}{r}\left[\frac{Ea + (.....).......}{r}\right]}$$

Q.21 To test the homogeneity of error variance obtained from family to family, the test applicable is known as "____________ of homogeneity of variances.

15.3 Multiple Choice Questions and Answers

Q.1 The replicated experiment designed to compare a number of parent plants w.r. to some attribute under study is known as

(a) Progeny row trial
(b) Compact family block design
(c) Split plot design
(d) Randomised complete block design

Q.2 The mass selection technique was adopted to select the plants showing superior genetic values in

(a) Plant breeding experiment
(b) Agronomy's experiment
(c) Crop cutting experiment
(d) Experiment on plant protection

Q.3 'A progeny row trial' adopted as an improved technique over mass selection technique was investigated by

(a) Agronomists (b) Plant breeders
(c) Entomologists (d) Crop physiologists

Q.4 In a progeny row trial each plot consists of a single row and all the seeds of the row belong to the

(a) Same parent plant
(b) Progeny plant
(c) Different parent plants
(d) Different plant of F2 generation

Q.5 Let there be p progenies and r replications with ki plants in each plot. Then d. f for progeny is equal to

(a) (r -1) (b) (p -1)
(c) r (p - 1) (d) p (r - 1)

Q.6 A progeny row trial is carried out in a simple

(a) C.R.D. (b) R.B.D.
(c) L.S.D. (d) Compact family block design

Q.7 In Question 5, the error (plot error) d.f. is equal to

(a) (ki - 1) (b) r (p - 1)
(c) (r - 1) (p - 1) (d) p (r - 1)

Q.8 In a layout described in Question 5 the d.f. for pooled plant error is equal to

(a) $r\sum_{i}^{rp}(ki-1)$ (b) $\sum_{i}^{rp}(ki-1)$

(c) $\sum_{i}^{rp}p\,(ki-1)$ (d) (r - 1) (p -1)

Q.9 A progeny row trial has been found suitable for the crop

(a) Wheat (b) Cotton

(c) Sorghum (d) All the three crop

Q.10 If the progenies tried in an experiment belong to a number of families then the type of comparison made in

(a) The comparison between different families

(b) Comparison between progenies within family

(c) Comparison between families within progenies

(d) Both (a) and (b)

Q.11 A layout suitable for studies as considered in Question 10, is known as

(a) R.B.D.

(b) Split plot design

(c) Progeny row trial

(d) Compact family block design

Q.12 In a compact family block design the progenies belonging to the same family are sown side by side in a family

(a) Main plot (b) Sub plot

(c) Row (d) Block

Q.13 In compact family block design, the family plots are randomised in

(a) Each main plot (b) Each block

(c) Each sub plot (d) Each row

Q.14 The progeny plots are randomised within the

(a) Main plots (b) Family plots

(c) Sub plots (d) Both (a) and (b)

Q.15 Compact family block design is analogous to the

(a) R.C.B.D. (b) L.S.D.

(c) Split plot design (d) Strip plot design

Q.16 For f families, p progenies in each family and r replications, in compact family block design the error d.f for main plot is equal to

(a) (r - 1) (p - 1) (b) (r - 1) (f - 1)

(c) (p - 1) (f - 1) (d) p(r - 1) (f - 1)

Q.17 In compact family block design as discussed in Question 16, the d.f for progenies within family is equal to

(a) p (f - 1) (b) f (p - 1)

(c) r (p - 1) (d) r (f - 1)

Q.18 As described in Question 16, the d.f. for sub plot error is equal to

(a) (fpr - 1) (b) r (p - 1) (f - 1)

(c) p (r - 1) (f - 1) (d) f (r - 1) (p - 1)

Q.19 In compact family block design, the subplot treatments are

(a) Different for all the main plots

(b) Same for all the main plots

(c) Same for all the blocks

(d) Different for all the blocks

Q.20 S.E. (Difference between two progeny means taken from two different families) is equal to

(a) $\sqrt{\frac{2}{r}\left[\frac{E_a+(p+1)E_b}{r}\right]}$ (b) $\sqrt{2\left[\frac{E_a+(p+1)E_b}{r^2}\right]}$

(b) $\sqrt{\frac{r}{2}\left[\frac{E_a+(p+1)E_b}{p}\right]}$ (d) $\sqrt{\frac{2}{r}\left[\frac{E_a+(p+1)E_b}{p}\right]}$

Q.21 To test the homogeneity of error variances obtained from family to family, the appropriate test has been suggested by

(a) Prof. R.A. Fisher (b) Prof Kemthorn

(c) Prof. Bartlett (d) Prof R.C. Mahalanobis

15.4 Key Answers

15.4.1 True / False

1.	True	2.	True	3.	True
4.	True	5.	True	6.	False
7.	True	8.	True	9.	True
10.	True	11.	True	12.	True
13.	True	14.	True	15.	True
16.	True	17.	True	18.	True
19.	True	20.	True		

15.4.2 Fill in The Blanks

1. Progeny row trial
2. Superior genetic
3. Mass selection
4. Single row, Parent plant
5. R. B. D, Progeny row, Each block
6. (p -1)
7. (r -1) (p - 1)
8. $\sum_{i}^{rp}(ki-1)$
9. Two type of
10. Families , Within
11. Compact family block design
12. Same family, main
13. Randomised
14. Within, sub plots
15. Split plot design
16. (r- 1) (f - 1)
17. f (p - 1)
18. (rpf - 1)
19. Different
19. (q - 1) Ec

20. $\sqrt{\left[\frac{(p+1)\,Eb}{r}\right]}$

21. Bartlett test

15.4.3 Multiple Choice Questions and Answers

1.	a	2.	a	3.	b
4.	a	5.	b	6.	b
7.	c	8.	b	9.	d
10.	d	11.	d	12.	a
13.	b	14.	b	15.	c
16.	b	17.	b	18.	a
19.	a	20.	a	21.	c

16

Analysis of Covariance (ANCOVA)

16.1 True / False

Q.1 Analysis of covariance is an extension of ANOVA.

Q.2 In an ANCOVA technique, observations are taken on more than one variable from each experimental unit.

Q.3 The main interest of an ANCOVA is to test whether the variation in the dependent variable y over the classes is due to class effects or due to its "dependence on independent variate x which also varies from class to class.

Q.4 An ANCOVA controls the experimental error by taking into consideration of dependence of y on x.

Q.5 An ANCOVA technique does not increase the precision of a randomized test experiment.

Q.6 An ANCOVA technique is not useful to fit regression in a multiple classification.

Q.7 The treatment and regression effects may not be additive for valid use of ANCOVA technique.

Q.8 In ANCOVA, it is assumed that the correct form of regression line has been footed.

Q.9 In ANCOVA, the danger of misleading result is greater when x shows real difference from treatment to treatment.

Q.10 The normality assumption of the error term in ANOVA in fixed effect model is needed only for hypothesis testing and interval estimation.

Q.11 The point estimaters and their estimated variances remain valid ever when normality does not hold.

Q.12 Heterosedasticity and correlation of errors do not create bias in the estimators.

Q.13 For random effect model and mixed effect model, the estimators of variance components remain unbiased even with non-normal random effects.

Q.14 In the words of Prof. R.A Fisher "ANCOVA combines the advantage and reconciles the requirements of the two very applicable procedurs known as regression and ANOVA".

Q.15 Estimation of variation between blocks or between rows or between columns from the residuals variation is possible by using the technique of "Local Control".

Q.16 The theory of ANCOVA makes use of concepts of both "ANOVA" and Regression Analysis".

Q.17 The theory of ANCOVA can not be applied to partition the total "Covariance" or "Cross Product" of two variablas y and x into its components.

Q.18 The theory of ANCOVA can be utilised to make an adjustment in mean of the dependent variate y for the differences in the values of independent (covariate) variate x.

Q.19 The variation σ^2 among observations taken on experimental units can be reduced by using one or more covariates in an ANCOVA.

Q.20 The assumptions for ANCOVA, are the combinations of assumptions of "ANOVA" and "Regression Analysis".

Q.21 In usual notations, the statistical model for ANCOVA in R.C.B.D. with one observation/cell can be written as $y_{ij} = \mu + t_i + b_j + \beta(x_{ij} - \bar{x}..)e_{ij}$

Q.22 The statistical model written in Question 21, can also be written as

$$y_{ij} - \beta(x_{ij} - \bar{x}..) = \mu + t_i + b_j + e_{ij}$$

Q.23 The model written in Q.22 is the ANOVA model suitable in R.C.B.D for the dependent variate Y adjusted for the effect of an independent variate x.

Q.24 The statistical model written in Q.21, can also be written as

$$y_{ij} - t_i - b_j = \mu + \beta(x_{ij} - \bar{x}..) + e_{ij}$$

Q.25 In statistical model written in Q.24, the stress has been given to measure the regression of y on x without the interference of treatment and block effect on y.

Q.26 In the theory of ANCOVA, the assumptions made on covariate x that the x's are fixed, measured without errors, and are independent of treatment effects.

Q.27 The test statistic to test the null hypothesis

Ho: No. differences among unadjusted treatment mean for y variate is defined as.

$$F = \frac{Tyy/(t-1)}{Eyy/(r-1)(t-1)} \sim {}^{F}(t-1),(r-1)(t-1)$$

Q.28 The regression coefficient b_{yx} of y on x for experimental error to make adjustement in treatment mean of y variate is defined as

$b_{yx} = E_{xy} / E_{xx}$

Q.29 The E.S S of y (ie Eyy) assignable to regression on x is defined as

$$E_{yy} = b_{yx}\, E_{xy} = \frac{(E_{yx})^2}{E_{xx}}$$

where $Exy = Eyx = \sum_{ij} (x_{ij} - \overline{x}..)(y_{ij} - \overline{y}..) - R_{xy} - T_{xy}$

Q.30 Adjusted Error SS for y variate can be defined as

$$\text{adjusted } Eyy = E_{yy} - \frac{(E_{xy})^2}{E_{xx}}$$

Q.31 The d.f for adjusted Ey*y = (r - 1)(t -1) -1

Q.32 The adjusted residual (ie error) M S denoted by : $s^2y.x$ is defined as

$$s^2y.x = \frac{Adj\, Eyy}{(r-1)(t-1)-1}$$

Q.33 The (treatment + Error) S.S. for y variate is defined as

$$= Syy - \frac{(Sxy)^2}{Sxx}$$

Q.34 The d.f. for (Treatment Error) SS(Y) = r (t - 1) -1

Q.35 $\text{Adjusted } Tyy = \left[Syy - \frac{(Sxy)^2}{Sxx}\right] - \left[Eyy - \frac{(Exy)^2}{Exx}\right]$

Q.36 To estimate the variance within each and all population of Y values of y, the appropriate error M. S. used is -

$$S^2_{y.x} \frac{Adj\, Eyy}{(r-1)(t-1)-1}$$

Q.37 If all the treatment $\overline{x}$'s are assumed to be equal then estimate of expected SS for two combined sources (treatment and error) is equal to -
Adjusted S S (Treatment + error)

Q.38 The homogeneity of linear regression (byx) of yon x for all treatments for block differences is necessary for valid and useful adjustment.

Q.39 Homogeneity of regression coefficients (byx) for applying an ANCOVA technique is an assumption just as is the assumption of homogeneity error variance in case for applying an ANOVA technique.

Q.40 To test the null hypothesis
Ho: There is no difference among the treatment means for the variate y adjusted for the regression of y on x, the F statistic is defined as

$$F = \frac{\text{M.S (Adjusted treatment mean for y)}}{\text{Adjusted } S^2 y.x}$$

$\sim F(t - 1), (r - 1)(t - 1) - 1$

Q.41 The regression equation to work out the adjusted treatment means of the variate y is written as-
Adjusted $\hat{Y}i = \hat{Y}i - \text{byx}(\overline{x}_i - \overline{x}..)$
where byx= b is the regression coefficient as defined in Q.28

Q.42 The result $\sum \text{byx}(\overline{x}_i - \overline{x}..) = 0$, hold good because byx for all treatment are same as defined byx $= \frac{Eyx}{Exx}$

Q.43 S.E. (Adjusted treatment, $\hat{Y}_i$) = $s_{\hat{Y}i} = sy.x\sqrt{\frac{1}{r} + \frac{(\overline{x}i. - \overline{x}..)^2}{Exx}}$

Q.44 S.E (Diff. bet. two Adj. Treatment) i e

$$S.E.(\hat{Y}_i - \hat{Y}_i'.) = sy.x\sqrt{\frac{2}{r} + \frac{(\overline{x}_i. - \overline{x}..)^2}{Exx}}$$

Q.45 The appropriate value of S.E. $(\hat{Y}_i - \hat{Y}_i.)$ as suggested by Finny is equal to = sy.x $\sqrt{\frac{2}{r}\left(1 + \frac{Txx}{(t-1)Exx}\right)}$

16.2 Fill in the Blanks

Q.1 Analysis of covariance is an extension of ____________

Q.2 In an ANCOVA technique observations are taken on more than one variable on ____________ units.

Q.3 The main interest of an ANCOVA is to test whether the variation in the dependent variable y over the classes due to class effects or due to its dependence on ____________

Q.4 An ANCOVA controls the experimental error by taking into considerations of dependence of y on ____________

Q.5 An ANCOVA techniqe ____________ the precision of a randomised test experiment.

Q.6 In ANCOVA technique is useful to fit ____________ in a multiple classification.

Q.7 In ANCOVA, it is assumed that the correct form of ____________ has been footed.

Q.8 The assumption is that the treatment and regression effects may not be additive for valid use of an ____________ technique.

Q.9 In ANCOVA, the danger of misleading result is greater when x shows real difference from ____________ to

Q.10 The normality assumption of the error term in ANOVA's fixed effect model is needed only for ____________ testing and ____________ estimation.

Q.11 The point estimators and their estimated variances remains valid even when ____________does not hold.

Q.12 Heterosedasticity and correlation of errors do not create ____________ in the estimators.

Q.13 For random effect model and mixed effect model, the estimators of variance components remain ____________ even with ____________ random effects.

Q.14 Estimation of variation between blocks or between rows or between columns from the residual variation is possible by using the technique of "____________"

Q.15 The theory of ANCOVA makes use of concepts of both "______________ and ______________"

Q.16 The theory of ANCOVA can be applied to partition the total "______________ or ______________ product" of two variables y and x into its components.

Q.17 The theory of ANCOVA can be utilised to make an adjustment in mean of the ______________ variable for the difference in the values of ______________ variable.

Q.18 The variance σ_y^2 among observations taken on experimental units can be reduced by using one or more ______________ in an ANCOVA.

Q.19 The assumptions for ______________ are the combinations of assumptions of "______________" and "Regression Analysis"

Q.20 In usual notations, the statistical model for ANCOVA in R.C.B.D. with one observaion/ cell can be written as-

$$y_{ij} = \mu + t_i + b_j + \ldots\ldots\ldots\ldots + e_{ij}$$

Q.21 The statistical model written in Question 20 can also be written as

$$y_{ij} = \beta(\ldots\ldots\ldots) = \mu + t_i + \ldots\ldots\ldots\ldots + e_{ij}$$

Q.22 The model mentioned in Question 21, is the ANOVA model suitable in R.C.B.D. for the ______________ variate y adjusted for the effect of an ______________ variate x

Q.23 The statistical model mentioned in Question 20 can also be written as

$$y_{ij} - t_i - b_j = \mu + \beta(\ldots\ldots\ldots\ldots) + e_{ij}$$

Q.24 In statistical model written in Question 20 the stress has been given to measure the regression of ______________ on x without the interference of ______________ and effects on y.

Q.25 In the theory of ANCOVA, the assumptions made on covariate x is that the x's are fixed, measured ______________ errors and are ______________ of treatment effects.

Q.26 The test statistic to test the null hypothesis

Ho: No difference among unadjusted treatment mean for y variate is defined as

$$F = \frac{Tyy/(t-1)}{Eyy/(\ldots\ldots\ldots)(\ldots\ldots\ldots)} \sim {}^{F}(t-1),(\ldots\ldots)(\ldots\ldots)$$

Q.27 The ____________ coefficient byx of y on x for experimental error to make adjustment in treatment mean of ____________ variate y is defined as

$$\text{byx} = \frac{\text{.........}}{\text{.........}}$$

Q.28 The error S.S of y (ie Eyy) assignable to regression on x is defined as

$$\text{Eyy} = \text{\_\_\_\_\_\_\_\_\_\_\_\_} \quad \text{Exy} = \frac{(\text{.......})^2}{\text{Exx}}$$

$$\text{where Exy} = \text{Eyx} = \sum(xi - \bar{x}..)(y_{ij} - \bar{y}..) - R_{xy} - T_{xy}$$

Q.29 Adjusted error S S for Y variate can be defined as

$$\text{Adjusted Eyy} = \text{Eyy} - \frac{(\text{......})^2}{\text{Exx}}$$

Q.30 The d.f. for Adjusted Eyy is equal to (r - 1) (.......)-1

Q.31 The adjusted residual (error) M.S denoted by s^2y.x is defined as

$$s^2\text{y.x} = \frac{\text{Adj Eyy}}{(\text{......})(\text{........}) - 1}$$

Q.32 The (Treatment + Error) SS for y is defined as-

$$\text{(Treat + Error) SS (y)} = \text{Syy} - \frac{(\quad)^2}{\text{Sxx}}$$

Q.33 The d.f for (Treat + Error) S.S (y) = ____________

Q.34 $$\text{Adjusted Tyy} = \left[\text{Syy} - \frac{(\text{Sxy})^2}{\text{Sxx}}\right] - \left[\text{Eyy} - \frac{\text{.........}}{\text{.........}}\right]$$

Q.35 To estimate the variance within each and all population values' of y the appropriate error M.S. is defined as

$$s^2\text{y.s} = \frac{\text{Adj Eyy}}{(\text{......})(\text{.......}) - 1}$$

Q.36 If all the treatment $\bar{X}$'s are assumed to be equal then estimate of expected S.S. for two combined sources (Treatment and Error) is equal to ____________

S.S. (Treat + Error)

Q.37 The _____________ of linear regression coefficient (byx) of y on x for all treatment adjusted _____________ is necessary for valid and useful adjustment.

Q.38 Homogeneity of regression coefficient (byx) to apply an _____________ technique b an assumption just as is the assumption of homogeneity of error variances to apply on _____________ technique.

Q.39 To test the null hypothesis-

Ho: No differences among adjusted treatment means, the F statistic is defined as

$$F = \frac{MS(\text{Adjusted treat.mean.for y})}{\text{Adjusted } s^2_{y.x}} \sim F\,(....),(....)(....)$$

Q.40 The regression equation to work out the adjusted treatment mean is written as-

$\text{Adj } \hat{y}_i = y_i - (......)(......)$

Q.41 The result $\sum_i byx(\overline{x}i - \overline{x}..) = 0$ holds good because byx for all treatments are _____________ ie byx is constant

Q.42 S.E. (Adj Treat. means $\hat{y}_{i.}$) = $s\hat{\overline{y}}_{i.} = \sqrt{\frac{1}{....} + \frac{(.........)^2}{Exx}}$

Q.43 S.E. (Diff. bet. treat mean) i e S. E. ($\hat{\overline{y}}_i - \hat{\overline{y}}_{i'.}$) is equal to

$$sy.x\sqrt{\frac{.....}{r} + \frac{(\quad)^2}{Exx}}$$

Q.44 The appropriate value of S.E. $(y_i - y_i{}'.)$ as suggested by Finny is equal to = $sy.x\sqrt{\frac{2}{r}\left(1 + \frac{Txx}{(.....)....}\right)}$

16.3 Multiple Choice Questions and Answers

Q.1 Analysis of covariance is an extension of _____________

(a) Student't test (b) Variance ratio (F) test

(c) Analysis of variance technique (d) Chisquare test

Q.2 In an ANCOVA, observations are taken on more than one variable on each ______________

(a) Experimental unit (b) Main plot unit
(c) Sub plot unit (d) Experimental material

Q.3 The main interest of an ANCOVA is to test whether the variation in the dependent variable y is due to its dependence on other variable x known as ______________

(a) Covariate (b) Auxiliary variate
(c) Regressor variate (d) All the (a), (b) and (c)

Q.4 An ANCOA controls the experimental error by taking into consideration of

(a) Dependence of the variate y on variate x
(b) Suitable experimental design adopted
(c) Randomisation Technique followed
(d) Both (a) and (b)

Q.5 The precision of a randomised test experiment can be increased by adopting

(a) An ANOVA technique
(b) Any experimental design
(c) Split plot design
(d) Compact family block design

Q.6 An ANCOVA technique is useful to fit regression in

(a) Two way classification (b) Multiple classification
(c) One way classification (d) Both (a) and (b)

Q.7 For a valid use of ANCOVA technique, the treatment and regression effects

(a) May not be additive (b) Must be additive
(c) Must be multiplicative (d) Must be exponential

Q.8 In ANCOVA technique, it is assumed that the correct form of the regression line

(a) Has not been footed (b) Has been footed
(c) Can not been established (d) None is true

Q.9 In ANCOVA, the danger of misleading result is greater when x _______

(a) Shows real difference among treatment

(b) Shows no real difference among treatment

(c) Depends on the variate Y

(d) Is not related with y

Q.10 The normality assumption of the error term in ANOVA in case of fixed effect model is needed only for the

(a) Hypothesis testing and interval estimation

(b) Hypothesis testing and point estimation

(c) Point and Interval estimation

(d) Interval estimation

Q.11 The point estimators and their estimated variances remain valid

(a) When normality holds for error term

(b) Even when normality does not hold for error term

(c) When errors are randomly distributed

(d) All the three (a) (b) and (c) are correct

Q.12 Heterosedastisity and correlation of error

(a) Do not make the estimator biased

(b) Make the estimator unbiased

(c) Make the amont of bias of the estimators equal to zero

(d) All the three (a), (b) and (c) are true

Q.13 In both cases of random effect and mixed effect model, the estimators of variance components remain

(a) Unbiased with normal and random effects

(b) Unbiased even with non-normal and random effects

(c) Biased with normal and non random effects

(d) Both (a) and (b) are true

Q.14 In the words of prof. R. A. Fisher " ANCOVA combines the advantage and reconciles the requirements of two very applicable procedures known as

(a) ANOVA and regression

(b) ANOVA and M.L.E. procedure

(c) Regression and principle of least square

(d) Both (a) and (b)

Q.15 Estimation of variation between blocks or between rows or between columns from residual variation is possible by using the technique of

(a) Randomisation (b) Local control

(c) Least square theory (d) Both (a) and (c)

Q.16 The total "covariance" or "cross product" of two variates y and x can be partitioned into its component using the

(a) ANCOVA technique

(b) Using ANOVA technique

(c) Using ANOVA and Regression analysis

(d) By both (a) and (c)

Q.17 The theory of ANCOVA makes use of concepts of both technique of

(a) ANOVA and Principle least square

(b) ANOVA and Regression analysis

(c) ANOVA and Sub plot

(d) Regression and Sub plot

Q.18 The theory of ANCOVA can be utilised to make on adjustment in mean of y for the difference in the values of.

(a) Covariate x

(b) Independent variate x

(c) Regression variable x

(d) All the three (a), (b) and (c)

Q.19 The variation σ_y^2 among observation taken y on experimental units can be reduced by using one or more covariaties in

(a) ANOVA (b) ANCOVA

(c) Regression Analysis (d) Split plot technique

Q.20 Assumptions of ANCOVA, are the combined assumptions of

(a) ANOVA and Regression Analysis

(b) ANOVA and Split Plot

(c) Regression and Sub Plots

(d) Both (b) and (c)

Q.21 In usual notations, the statistical model for ANCOVA in R.C. B.D. with one observation/cell can be written as

(a) $y_{ij} = \mu + t_i + \beta(x_i - \overline{x}..) + e_{ij}$

(b) $y_{ij} = \mu + b_i + \beta(x_{ij} - \overline{x}..) + e_{ij}$

(c) $y_{ij} = \mu + t_i + bj + \beta(x_{ij} - \bar{x}..) + e_{ij}$
(d) $y_{ij} = \mu_i + t_i + bj + \beta(x_{ij} - \bar{x}..) + e_{ij}$

Q.22 The correct statistical model written is Q.21, can also be written as
(a) $y_{ij} = \beta(xi - \bar{x}..) = \mu + t_i + + e_{ij}$
(b) $y_{ij} = \beta(xi - \bar{x}..) = \mu + b_j + e_{ij}$
(c) $y_{ij} = \beta(xi - \bar{x}..) = \mu_i + t_i + e_{ij}$
(d) $y_{ij} - \beta(xi - \bar{x}..) = \mu_i + t_i + b_j + e_{ij}$

Q.23 The model written in Question 22 (d) is the ANOVA model in R.C.B.D., of the variate y adjusted for the effect of
(a) Covariate x
(b) Auxiliary variate x
(c) Independent variate
(d) For all three (a), (b) and (c)

Q.24 The statistical model written in Question 21 (c) can also be written as
(a) $y_{ij} - t_i - b_j = \mu + \beta(x_{ij} - \bar{x}..) + eij$
(b) $y_{ij} - b_j = \mu + \beta(x_{ij} - \bar{x}..) + eij$
(c) $y_{ij} - t_i = \mu + \beta(x_{ij} - \bar{x}..) + eij$
(d) $y_{ij} - t_i - b_j = \mu_i + \beta(x_{ij} - \bar{x}..) + e_{ij}$

Q.25 In statistical model in Question 24 (a), the stress has been given to measure the regression of y on x without the interference of
(a) Treatment effect on y
(b) Block effect on y
(c) Treatment and block effects on y
(d) None of the three

Q.26 In the theory of ANOVA, the assumptions made on covarite x is that x's are
(a) Fixed,
(b) Measured without error
(c) Are independent of treatment effect
(d) The combination of (a), (b) and (c)

Q.27 To test the null hypothesis
Ho: "There are no difference among unadjusted treatment means of y variate", the test statistic F is defined and distributed as

(a) $F = \dfrac{Tyy/(t-1)}{Eyy/(r-1)(t-1)} \sim {}^{F}(t-1),(r-1)(t-1)$

(b) $F = \dfrac{Tyy/(r-1)}{Eyy/(t-1)(r-1)} \sim {}^{F}(t-1),(r-1)(t-1)$

(c) $F = \dfrac{Tyy/(r-1)}{Exx/(t-1)(r-1)} \sim {}^{F}(t-1),(r-1)(t-1)$

(d) $F = \dfrac{Txx/(t-1)}{Exx/(t-1)(r-1)} \sim {}^{F}(t-1),(r-1)(t-1)$

Q.28 The regression coefficient by x of y on x for experimental error to make the adjustement in treatment mean for y variate is defined as

(a) byx = Eyy /Exx

(b) $byx = \dfrac{Exy}{Eyy}$

(c) $byx = \dfrac{Exy}{Exx}$

(d) $byx = \dfrac{Exy}{Eyy}$

Q.29 The error S.S of y (ie Eyy) assignable to regression on x is

(a) Eyy = byx Exy,

(b) $Eyy = \dfrac{(Eyx)^2}{Exx}$

(c) $Eyy = \dfrac{Exy}{Exx}$

(d) Both (a) and (b)

Q.30 Adjusted error S.S (y) can be defined as

(a) $\text{Adj Eyy} = Eyy - \dfrac{(Exy)^2}{Exx}$

(b) $\text{Adj Eyy} = Eyy - \dfrac{Exy}{Exx}$

(c) Adj Eyy = Eyy - byx Exy,

(d) By both (a) and (c)

Q.31 The d.f for adjusted Eyy is euqal to

(a) (r - 1) (t - 1) -1

(b) (r - 1) (t - 1) +1

(c) (r - 1) (t - 1) -2

(d) (r - 1) (t - 1) +2

Q.32 The adjusted residual M.S. denoted by $s_y^2.x$ is defined as-

(a) $s_y^2.x = \frac{\text{Adj Eyy}}{(r-1)(t-1)+1}$ (b) $s_y^2.x = \frac{\text{Adj Eyy}}{(r-1)(t-1)-1}$

(c) $s_y^2.x = \frac{\text{Adj Eyy}}{(r-1)(t-1)-2}$ (d) $s_y^2.x = \frac{\text{Adj Eyy}}{(r-1)(t-1)+2}$

Q.33 The (Treatment + Error) S. S for y variate is defined as

(a) $\text{Syy} - \frac{(\text{Exy})^2}{\text{Exx}}$ (b) Syy - byx Sxy,

(c) $\text{Syy} - \frac{\text{Exy}}{\text{Exx}}$ (d) Both (a) and (b)

Q.34 The d.f. for (Treat + Error) S.S (Y) is equal to

(a) r(t - 1)-1 (b) r(t - 1)+1

(c) rt - r -1 (d) Both (a) and (c)

Q.35 Adjusted Tyy is equal to

(a) $\left[\text{Syy} - \frac{(\text{Sxy})^2}{\text{Sxx}}\right] - \left[\text{Eyy} - \frac{(\text{Exy})^2}{\text{Exx}}\right]$

(b) $\left[\text{Syy} - \frac{(\text{Sxy})^2}{\text{Sxx}}\right] - [\text{Eyy} - \text{byx Exy}]$

(c) $[\text{Syy} - \text{byx Sxy}] - [\text{Eyy} - \text{byx Exy}]$

(d) All the (a) (b) and (c) are correct

Q.36 To estimate the variance within each and all population values, of y the appropriate error M.S. used as $s_{y.x}^2$ is equal to

(a) $\frac{A_{dj}\text{ Eyy}}{(r-1)(t-1)-1}$ (b) $\frac{A_{dj}\text{ Eyy}}{(r-1)(t-1)+1}$

(c) $\frac{A_{dj}\text{ Eyy}}{(r-1)(t-1)+2}$ (d) $\frac{A_{dj}\text{ Eyy}}{(r-1)(t-1)-2}$

Q.37 The homogeneity of by x to apply ANCOVA is an assumption of homogeneity of error variance to apply

(a) ANOVA technique (b) Regression Analysis
(c) Chi-square test (d) Bartlett test

Q.38 To test the null hypothesis

Ho : There is no difference among adjusted treatment means $(\overline{y}i.)$, the F statistic is defined and distributed as

(a) $F = \frac{MS(\text{Adjusted } \overline{y}i.)}{\text{Adjusted } s^2 y.x} \sim {}^{F}(t-1),(r-1)(t-1)-1$

(b) $F = \frac{MS(\text{Adj } \overline{y}i.)}{\text{Adj } s^2 y.x} \sim {}^{F}(r-1),(t-1)(r-1)-1$

(c) $F = \frac{M.S(\text{Adj } \overline{y}i.)}{\text{Adj } s^2 x.y} \sim {}^{F}(t-1),(t-1)(r-1)-2$

(d) $F = \frac{M.S(\text{Adj } \overline{y}i.)}{\text{Adj } s^2 x.y} \sim {}^{F}(t-1),(t-1)(t-1)+1$

Q.39 The regression equation to work out the Adj $\overline{y}i$ is written as-

(a) Adj $\hat{\overline{y}}i. = \overline{y}i. - \text{byx}(\overline{x}i. - \overline{x}..)$
(b) Adj $\hat{\overline{y}}i. = \overline{y}i. + \text{byx}(\overline{x}i. - \overline{x}..)$
(c) Adj $\hat{\overline{y}}i. = \overline{y}i. + \text{bxy}(\overline{x}i. - \overline{x}..)$
(d) Adj $\hat{\overline{y}}i. = \overline{y}i. - \text{bxy}(\overline{x}i. - \overline{x}..)$
Where byx = b is

Q.40 The result $\sum_i \text{byx}(\overline{x}i. - \overline{x}..) = 0$ hold good because-

(a) byx for all $\overline{y}i$ is different (b) byx for all $\overline{y}i$ is same
(c) bxy for all $\overline{y}i$ is different (d) bxy for all $\overline{y}i$ is same

Q.41 The S. E. (Adj mean $\hat{\overline{y}}i$) is equal to-

(a) $sy.x\sqrt{\frac{1}{r} + \frac{(xi - \overline{x}..)^2}{Exx}}$ (b) $sy.x\sqrt{\frac{2}{r} + \frac{(\overline{x}i - \overline{x}..)^2}{Exx}}$

(c) $sy.x\sqrt{\frac{1}{r} + \frac{(\overline{x}i - \overline{x}..)^2}{Eyy}}$ (d) $sy.x\sqrt{\frac{2}{r} + \frac{(\overline{x}i - \overline{x}..)^2}{Eyy}}$

Q.42 S.E. (Diff. Bet. two Adj mean) ie S.E $(\overline{y}_{i.} - \overline{y}_{i.})$ is equal to

(a) $sy.x\sqrt{\frac{2}{r}+\frac{(\overline{x}i.-\overline{x}..)^2}{Exx}}$ (b) $sy.x\sqrt{\frac{1}{r}+\frac{(\overline{x}i.-\overline{x}..)^2}{Exx}}$

(c) $sy.x\sqrt{\frac{2}{r}+\frac{(\overline{x}i.-\overline{x}..)^2}{Eyy}}$ (d) $sy.x\sqrt{\frac{1}{r}+\frac{(\overline{x}i.-\overline{x}..)^2}{Eyy}}$

Q.43 The appropriate value of S.E. $(\overline{y}_{i.} - \overline{y}_{i.})$ as suggested by Finny is equal to-

(a) $sy.x\sqrt{\frac{2}{r}\left(1+\frac{Txx}{(t-1)Exx}\right)}$

(b) $sy.x\sqrt{\frac{1}{r}\left(1+\frac{Txx}{(t-1)Exx}\right)}$

(c) $sy.x\sqrt{\frac{2}{r}\left(1+\frac{Txx}{(t+1)Exx}\right)}$

(d) $sy.x\sqrt{\frac{2}{r}\left(1+\frac{Tyy}{(t-1)Exx}\right)}$

16.4 Key Answers

16.4.1 True / False

1.	True	2.	True	3.	True
4.	True	5.	False	6.	False
7.	True	8.	True	9.	True
10.	True	11.	True	12.	True
13.	True	14.	True	15.	True
16.	True	17.	False	18.	True
19.	True	20.	True	21.	True
22.	True	23.	True	24.	True
25.	True	26.	True	27.	True

28.	True	29.	True	30.	True
31.	True	32.	True	33.	True
34.	True	35.	True	36.	True
37.	True	38.	True	39.	True
40.	True	41.	True	42.	True
43.	True	44.	True	45.	True

16.4.2 Fill in the Blanks

1. ANOVA
2. Each experimental
3. Covariate x
4. Covariate x
5. Increases
6. A regression line
7. Regression line
8. ANCOVA
9. Treatment, treatment
10. Hypothesis, Interval
11. Normality
12. Bias
13. Unbiased, Non-normal
14. Local Control
15. ANOVA, Regression Analysis
16. Covariance, Cross
17. Dependent, Independent
18. Covariates
19. ANOVA, ANOVA
20. $\beta\,(x_{ij} - \overline{x}..)$
21. $(x_{ij} - \overline{x}..)$, bj
22. Dependent, Independent
23. $(x_{ij} - \overline{x}..)$

24. Y, Treatment, Block
25. Without, Independent
26. (r - 1) (t - 1) , (r - 1) (t - 1)
27. Regression, Dependent $\frac{Exy}{Exx}$
28. byx, (Exy)
29. (Exy)
30. (t - 1)
31. (t - 1) (r - 1)
32. (Syx)
33. r (t - 1) -1
34. $\frac{(Exy)^2}{Exx}$
35. (r - 1) (t - 1)
36. Adjusted
37. Homogeneity, for block differences
38. ANCOVA , ANOVA
39. (t - 1), (r - 1) (t - 1)
40. byx $(\overline{x}_i. - \overline{x}..)$
41. Equal (same)
42. r, $(\overline{x}_i. - \overline{x}..)$
43. 2, $(\overline{x}_i. - \overline{x}_{i'}.)$
44. (t - 1) Exx

16.4.3. Multiple Choice Questions and Answers

1.	c	2.	a	3.	d
4.	d	5.	a	6.	d
7.	a	8.	b	9.	a
10.	a	11.	d	12.	d
13.	d	14.	a	15.	b
16.	d	17.	b	18.	d

19. b	20. a	21. c
22. d	23. d	24. a
25. c	26. d	27. a
28. c	29. d	30. d
31. a	32. b	33. d
34. d	35. d	36. a
37. a	38. a	39. a
40. b	41. a	42. a
43. a		

17

Some Miscellaneous Questions on ANOVA Technique

17.1 True / False

Q.1 In random effect model, the treatments are taken as a random sample of treatments drawn from a large group of T treatments.

Q.2 In varietal trials for plant breeding or genetics, the appropriate statistical model is not the random effect model.

Q.3 In a random effect model

$$y = \mu + T_i + e_{ij}$$

The effects μ & T_i are estimable independently.

Q.4 In Question 3, the effect ($\mu + T_i$) can be estimated.

Q.5 For (e_i & e_j) being two error terms on (i & j)th values in the model then covariance (ei, ej) $\neq$ 0

Q.6 When the deviations of the actual values y are taken from their estimated values, the residual S.S. will be minimum.

Q.7 In a random effect model $y = \mu + T_i + e_{ij}$, we are interested only in the variability of Ti.

Q.8 In a random effect model $y = \mu + T_i + e_{ij}$, we can not continue the experiment on a specified set of treatments

Q.9 In case of random effect model with n_i (Unequal) replication on i^{th} treatment we have

E (Treat M. S) $= \sigma^2 + n_o \sigma_r^2$

$$\text{Where } n_o = \frac{1}{(t-1)}\left[\sum_{i}^{t} n_i - \frac{\sum_{i}^{t} n_i^2}{\sum_{i-1} n_i}\right]$$

Q.10 In case of equal number of replication say r . for all treatment, we have E (Treat. M.S.) = $\sigma^2 + r\sigma_r^2$ with V (T) = σ_r^2

Q.11 In case of equal or unequal number of replications, we have E (Error M.S.) = σ^2, the population variance

Q.12 Let there be t treatments, each replicated r times, and s be the number of units (ie sample size) selected randomly from each experimental unit, then the statistical model in C. R. D will be written as

$$y_{ijk} = \mu + r_i + e_{ij} + \delta_{ijk}$$

Where δ_{ijk} is the value of sampling error arised on kth random unit drawn from $(i\,j)^{th}$ experiment unit.

Q.13 In Question 12, an experimental error eij will follow the normal distribution as

eij ~ N (o, σ_e^2) with constant error variance σ_e^2

Q.14 In Question 12, the sampling error term δ_{ijk} will be distributed as- δ_{ijk} ~ N (o, σ^2) where σ^2 is the population variance worked out on individual sampling unit (say, plants leaves)

Q.15 In Question 12, we have

E (Treat M.S.) = $\sigma^2 + s\sigma^2 + rs\ \sigma_t^2$

Q.16 Question 12 We have

E (Plots within treat) = Exptl. Error M.S. = $\sigma^2 + s\sigma_e^2$

Q.17 In Question 12 We have

E (Between units of sample within experimental units)

= E (sampling error M.S) = σ^2

Q.18 In Question 12, to test the significance of treatment means, the F statistic will be defined as

$$F = \frac{(\sigma^2 + s\sigma_e^2 + rs\sigma_r)^2}{(\sigma^2 + s\sigma_e^2)} \sim F(t-1), t(r-1)$$

Q.19 To test the significance of experimental error, the appropriate F statistic will be defined as-

$$F = \frac{(\sigma^2 + s\sigma_e^2)}{\sigma_e^2} \sim Ft(r-1), tr(s-1)$$

Q.20 If in Question 17, the F statistic is declared non-significant then the experimental error M.S and sampling error M.S. will be pooled to work out a common error M.S. and it will be utilised to test the significance of treatment means and block effects.

Q.21 As mentioned in Question 20, the pooled error M.S. is work out as

$$\text{Pooled Error (M.S)} = \frac{t(r-1)(\sigma^2 + s\sigma e^2) + tr(s-1)\sigma^2}{t(r-1) + tr(s-1)}$$

Q.22 As discussd in details in Question 12, then in case of R.C.B.D., the statistical model will be written as

$$y_{ijk} = \mu + T_i + \beta_j + e_{ij} + \delta_{ijk} \text{ with } \begin{pmatrix} i = 1,\ 2, \ldots..t \\ j = 1,\ 2, \ldots..r \\ k = 1, 2, \ldots..s \end{pmatrix}$$

Q.23 In statistical model for R.C.B.D., the expected M.S for block effect will be written as

E (Block M.S.) = $\sigma^2 + s\sigma_e{}^2 + ts\sigma_B 2$

Q.24 As in Question 23 We have

E (Treat M.S.) = $\sigma^2 + s\sigma_e{}^2 + rs\sigma_T 2$

Q.25 As in Question 23, the expected M.S. of experimental error is equal to E (Exptl. Error M.S.) = $\sigma^2 + s\sigma_e{}^2$

Q.26 As in Question 23, the expected M.S. of sampling error is given as

E (Sampling Error M.S.) = σ^2

Q.27 As details in Question 20, the experimental error d.f is equal to (r - 1) (t -1)

Q.28 As detail given in Question 20, the d.f. for the sampling error rt (s - 1)

Q.29 In Question 22, to test the significance of treatment and block effects, the error M.S. used to define F statistic will be "Experimental error M.S.

Q.30 To test the significance of experimental error eij; the test statistic will be defined as -

$$F = \frac{\text{Exptl. Error M.S.}}{\text{Sampling Error M.S.}} \sim {}^{F}(r-1)(t-1), rt(s-1)$$

Q.31 In case if variance of treatment are not homogeneous and we adopt an ANOVA technique then ANOVA applied will stand unvalid.

Q.32 If the homogeneity of variances (error variances) stands rejected then ANOVA technique can be applied on transformed data obtained through application of some suitable transformation on original data.

Q.33 The least significant difference or critical difference (C.D) methods is the most liberal test as compared to ANOVA technique.

Q.34 The L.S.D testor C.D. test slightly declares more number of pairs of treatments as significant than that which are actually significant.

Q.35 Duncan's multiple range test is slightly rigid or conservative test and is more popular in agriculture science.

Q.36 Duncan's multiple range test compares the range of any set of p means with an appropriate least significant range denoted by Rp and is defined as-

$$R_p = S_{\bar{x}}\, r_p$$

where $s_{\bar{x}}^2 = \hat{\sigma}_{\bar{x}}^2 = \frac{\sigma_e^2}{n}$ and σ_e^2 is the error variance.

Q.37 The value of r_p depends upon level of significance (a) and error d.f. ANOVA.

Q.38 The value of r_p may be obtained from a table known as "Critical value of Duncan's multiple range test" for $\alpha = 0.01$ and $p = 2, 3, \ldots\ldots\ldots, 10$ and error df from 1 to 120.

17.2 Fill in the Blanks

Q.1 In random __________, the treatment are taken as a random sample of t __________ drawn from a large group of T treatments.

Q.2 In varietal trials for plant breeding or genetics, the appropriate statistical model is the __________

Q.3 In random effect model

$$y_{ij} = \mu + t_i + e_{ij}$$

the effect _______ and __________ are not estimable in dependently.

Q.4 In Question 3 the effect (__________) can be estimated jointly

Q.5 For (ei and ej) being two error terms on (i, j)th value in the model then __________ (ei, ej) = 0

Q.6 When the deviations of the actual values y taken from their __________, the residual S.S will be __________

Q.7 In random effect model, $y = \mu + t_i + e_{ij}$, we are interested only in the __________ of t_i.

Q.8 In a random effect model $y = \mu + t_i + e_{ij}$, we can not continue the experiment on a specified set of __________

Q.9 In a random effect model, with n_i (unequal) replication on ith treatment, we have

E (Treat. M.S.) = σ^2_{+}..........

Where $n_o = \frac{1}{(t-1)}\left[\sum_{i=1}^{t} n_i - \sum_{i}^{t} n_i^2 / \sum^{t} ni\right]$

Q.10 In case of equal number of replication say r, for all treatment, we have

E (Treat M.S.) = __________ + σ_t^2 with V $(t_i) = \sigma_t^2$

Q.11 In case of equal or unequal number of replications of treatments, we have

E (Error M.S) = __________

Q.12 In case if s number of units are selected randomly from each experimental unit in C.R.D, the statistical model will be written as-

$y_{ijk} = \mu + ... + ... + \delta_{ijk}$

where ($y_{ijk,}$ μ, and δ_{ijk} have usual meaning)

Q.13 In Question 12, an experimental error eij will follow __________ distribution as eij ~ N (o, σ_e^2)

Q.14 In Question 12, the sampling error term δ_{ijj} ,will be distributed as δ_{ijj} ~ N(o, ______).

Q.15 In Question 12 we have E (Treat. M.S.) = σ_+^2 ______ + ______ σ_t^2

Q.16 In Question 12, We have

E (Plots M.S within treat = Exptl Error M.S) = __________

Q.17 In Question 12, we have E (Sampling error M.S.) = __________

Q.18 In Question 12, to test the significance of treatment means, the F statistic will be defined as

$$F = \frac{(\sigma^2 + s\sigma_e^{\ 2} + rs\sigma_t^{\ 2})}{(\ldots\ldots + \ldots\ldots\ldots)} \sim {}^{F}(t-1), t\,(r-1)(\ldots\ldots)$$

Q.19 To test the significance of experimental error, the appropriate F statistic will be defined as

$$F = \frac{(\ldots\ldots\ldots + \ldots\ldots\ldots)}{\sigma^2} \sim {}^{F}t\,(r-1), tr\,(\ldots\ldots\ldots)$$

Q.20 If experimental error $\sigma_e^{\ 2}$ is declared non significant then pooled error M.S will be worked out as

$$\text{Pooled error M.S} = \frac{t(r-1)(\ldots\ldots + s\sigma_e^{\ 2}) + tr\,(\ldots\ldots)\sigma^2}{t(r-1) + tr\,(s-1)}$$

Q.21 As discussed in Question 12, the statistical model for R.C.B.D., will be written as

$yijk = \mu +$ ______ + ______ + ______ $+ \ \delta_{ijk}$

Q.22 As discussed in (Question 12 and 21) jointly, the expected M.S. for block effect will be written as

$E(\text{Block M.S}) = \sigma^2 + \ldots\ldots\sigma e^2 + \ldots\ldots\ldots\sigma_\beta^{\ 2}$

Q.23 As in Question 22, we have

$E\ (\text{Treat. M.S}) = \sigma^2 + \ldots\ldots\sigma e^2 + \ldots\ldots\ldots\sigma_r^{\ 2}$

Q.24 As in Question 22, we have

$E\ (\text{Exptl. Error M.S.}) = \sigma^2 + s\ldots\ldots\ldots$

Q.25 As in Question 23, we have-

E (Sampling error M.S.)=

Q.26 As with details in Question 12 and 21, the error d.f for sampling error i.e.

d.f (Sampling error) = ______ (s - 1)

Q.27 As with details in Question 12 & 21, we have

d.f (Exptl error) = (______) (______)

Q.28 In Question 21, to test the significance of treatment and block effect, to define the F statistic, the error M.S applicable will be the " ______ "

Q.29 To test the significance of the experimental error eij, the test statistic F will be defined as

$$F = \frac{\text{Exptl Error M.S.}}{\text{Sampling Error M.S}} \sim {}^{F}(.)(.),\ldots..\ (s-1)$$

Q.30 If variances of treatments are declared significant and we adopt ANOVA technique, then the results obtained from ANOVA will stand ______

Q.31 If the homogeneity of error variances stands rejected then the ANOVA technique should be applied on _________ data, to get valid results.

Q.32 The least significant test or critical difference test is the most _________ as compared to ANOVA test.

Q.33 The L.S.D. test or test slightly declares ________ of pairs of treatments, as significant than that which are actually significant.

Q.34 Duncan's multiple range test is slightly _________ or conservative test and is more _________ in agricultural science.

Q.35 Duncan's multiple range test compares the range of any set of p _________ with an appropriates least significant range denoted by Rp which is defined as R_p = _________ r_p.

Q36 The value of r_p depends upon the level of _________ (α), and error _________ in ANOVA.

Q.37 The value of r_p may be obtained from a table known as " Critical value of _________ range test".

17.3 Multiple Choice Questions and Answers

Q.1 For a random effect model, t treatments under study, taken from a population of T treatment are considered as

(a) A random sample of size t.
(b) Systematic sample of size t.
(c) Simple random sample of size t.
(d) Both (a) and (c)

Q.2 For varietal trials in plant breeding or genetics, the appropriate statistical model applied is the

(a) Fixed effect model
(b) Mixed effect model
(c) Random effect model
(d) Both (a) and (b)

Q.3 In a random effect model, $y = \mu + r_i + e_{ij}$, the treatments

(a) μ and t_i are estimable separately and independently

(b) μ and t_i are not estimable separately and independently

(c) $(\mu + r_i)$ are estimable jointly

(d) Both (b) and (c)

Q.4 If (ei and ej) are two terms on two $(i, j)^{th}$ treatment in a statistical model then

(a) Cov (ei, ej) $\neq 0$

(b) Cov (ei, ej) = 0

(c) Cov (ei, ej) stands minimum,

(d) Cov (ei, ej) stands maximum

Q.5 For the deviations $(y_i - \hat{y}_i)$, the residual S.S will be

(a) Minimum

(b) Maximum

(c) Equal to zero

(d) Both (b) and (c)

Q.6 In a random effect model $y = \mu + t_i + e_{ij}$ we are interested in

(a) Estimating the effect of treatment t_i

(b) The variability of t_i

(c) The interval estimation for mean effect of t_i

(d) Both (b) and (c)

Q.7 In a random effect model $y = \mu + t_i + e_{ij}$, we cannot continue the experiment on a specified set of treatments (Ti , i= 1, 2 ____,) because.

(a) The set treatment is fixed

(b) The set treatment is a random sample

(c) The set treatment varies from time to time

(d) Both (b) and (c)

Q.8 In a random effect model with r_i replication on i^{th} treatment and ro $\left[\sum_{i=1}^{t} r_i - \sum_{i}^{t} r_i^2 / \sum_{i}^{t} r_i\right]$, we have E (Treat. M.S.) equal to

(a) $\sigma^2 + \text{ro}\,\sigma_t^2$

(b) $\sigma^2 + \left(\sum_{i}^{t} r_i\right)\sigma_t^2$

(b) $\sigma^2 + \frac{\sigma_t^2}{\text{ro}}$

(d) $\sigma^2 + \frac{\sigma_t^2}{\left(\sum_{i}^{t} r_i\right)}$

Q.9 In random effect model with same number of replication 'r', for all treatments, the value of E (Treat. M.S) will be equal to

(a) $\sigma^2 + \frac{\sigma_t^2}{r}$ (b) $\sigma^2 + r\sigma_t^2$

(c) $\sigma^2 + (r-1)\sigma_t^2$ (d) $\sigma^2 + (r-1)\sigma_t^2$

Q.10 In case of equal or unequal number of replication in ANOVA, the value of E (Error M.S.) will

(a) Be equal to error variance σ_e^2

(b) Be equal to population variance σ^2

(c) Not be equal to the population variance σ^2

(d) Both (a) and (c)

Q.11 Let the number of treatment and replications be (t, r) in a C.R.D. and s be the sample size drawn randomly from each experimental unit then the statistical model is written as

(a) $y_{ijk} = \mu + t_i + e_{ij}$ (b) $y_{ijk} = \mu + t_i + e_{ij} + \delta_{ijk}$

(c) $y_{ijk} = \mu + t_i + \delta_{ijk}$ (d) $y_{ijk} = \mu + e_{ij} + \delta_{ijk}$

Q.12 In Question 11, experimental eij will be normally distributed as

(a) eij ~ N (0, σ^2) (b) eij ~ N (0, σ_e^2)

(c) eij ~ N (σ_e^2, 0) (d) eij ~ N (σ^2, 0)

Q.13 In Question 12, the sampling error δ_{ijk} will be normally distributed as

(a) $\delta_{ijk} \sim N(0, \sigma_e^2)$ (b) $\delta_{ijk} \sim N(\sigma^2_{,}\ 0)$

(c) $\delta_{ijk} \sim N(0, \sigma^2)$ (d) $\delta_{ijk} \sim N(\sigma_e^2, 0)$

Q.14 In Question 11, we have E (Treat M.S.) will be equal to

(a) $\sigma^2 + s\sigma_e^2 + \sigma_r^2$ (b) $\sigma^2 + \sigma_e^2 + rs\,\sigma_r^2$

(c) $\sigma^2 + s\sigma_e^2 + r\sigma_r^2$ (d) $\sigma^2 + s\sigma_e^2 + rs\,\sigma_t^2$

Q.15 In Question 11, the E (Exptl Error M.S) will be equal to

(a) $\sigma_T^2 + s\sigma_e^2$ (b) $\sigma +^2 r\sigma_e^2$

(c) $\sigma_e^2 + r\sigma^2$ (d) $\sigma^2 + s\sigma_e^2$

Q.16 In Question 11, we have E (sampling M.S.) equal to

(a) σ^2 (b) $\sigma_e^{\ 2}$

(c) $\sigma^2 + s\sigma_e^{\ 2}$ (d) $\sigma^2 + r\sigma_e^{\ 2}$

Q.17 In Question 11 to test the significance of treatment means, the F is defined as

(a) $F = \dfrac{(\sigma^2 + s\sigma_e^{\ 2} + rs\,\sigma_T^{\ 2})}{(\sigma^2 + \sigma_e^{\ 2})} \sim {}^F(t-1),\, t\,(r-1)$

(b) $F = \dfrac{(\sigma^2 + r\sigma_e^{\ 2} + rs\,\sigma_T^{\ 2})}{(\sigma^2 + s\sigma_e^{\ 2})} \sim {}^F(r-1),\, t\,(r-1)$

(c) $F = \dfrac{(\sigma^2 + s\sigma_e^{\ 2} + rs\,\sigma_T^{\ 2})}{(\sigma^2 + s\sigma_e^{\ 2})} \sim {}^F(t-1),\, t\,(r-1)$

(d) $F = \dfrac{(\sigma^2 + \sigma_e^{\ 2} + \sigma_T^{\ 2})}{(\sigma^2 + \sigma^2)} \sim {}^F(t-1),\, r\,(t-1)$

Q.18 To test the significance of experimental error, the appropriate F statistic will be defined as

(a) $F = \dfrac{\sigma^2 + s\sigma_e^{\ 2}}{\sigma^2} \sim Ft\,(r-1),\, tr\,(s-1)$

(b) $F = \dfrac{\sigma^2 + s\sigma_e^{\ 2}}{\sigma^2} \sim Ft\,(t-1), (t-1)(r-1)$

(c) $F = \dfrac{\sigma^2 + r\sigma_e^{\ 2}}{\sigma^2} \sim F(r-1),(r-1)(t-1)$

(d) $F = \dfrac{\sigma^2 + s\sigma_e^{\ 2}}{\sigma^2} \sim {}^F(t-1),(t-1)(r-1)$

Q.19 The pooled common error M.S. of both "Exptl, Error M.S. and sampling Error (MS) will be utilised to test the significance of

(a) Treatment effect
(b) Block effect
(c) Neither treatment nor block effect
(d) Both (a) and (b)

Q.20 As illustrated in Question 11, the statistical model in case of R.C.B.D. will be written as

$$\begin{bmatrix} i = 1, 2, ..., t \\ j = 1, 2, ..., r \\ k = 1, 2, ..., s \end{bmatrix}$$

(a) $y_{ijk} = \mu + t_i + \beta_j + \delta_{ijk}$ (b) $y_{ijk} = \mu + t_i + \beta_j + e_{ij}$

(c) $y_{ijk} = \mu + \beta_j + e_{ij} + \delta_{ijk}$ (d) $y_{ijk} = \mu + t_i + \beta_j + e_{ij} + \delta_{ijk}$

Q.21 As illustrated in Question 20 the E (Block M.S.) will be equal to

(a) $\sigma^2 + s\sigma_e^2 + ts\,\sigma_B^2$ (b) $\sigma^2 + r\sigma_e^2 + \sigma_B^2$

(c) $\sigma^2 + s\sigma_e^2 + s\,\sigma_B^2$ (d) $\sigma^2 + t\sigma_e^2 + rt\,\sigma_B^2$

Q.22 In Question 21, the E (Treat M.S.) will be equal to

a) $\sigma^2 + s\sigma_e^2 + rs\,\sigma_T^2$ (b) $\sigma^2 + s\sigma_e^2 + ts\,\sigma_T^2$

(c) $\sigma^2 + t\sigma_e^2 + ts\,\sigma_T^2$ (d) $\sigma^2 + r\sigma_e^2 + rt\,\sigma_T^2$

Q.23 As illustrated in Question 20, E (Exptl Error M.S.) will be equal to

(a) $\sigma^2 + r\sigma_e^2$ (b) $\sigma^2 + t\sigma_e^2$

(c) $\sigma^2 + s\sigma_e^2$ (d) None

Q.24 As illustrated in Question 20, E (Sampling error M.S.) is equal to

(a) $\sigma^2 + s\sigma_e^2$ (b) $\sigma^2 + t\sigma_e^2$

(c) $\sigma^2 + r\sigma_e^2$ (d) σ^2

Q.25 As illustrated in Question 20, the d.f. for experimental error is equal to

(a) (r - 1) (s -1) (b) (t - 1) (s - 1)

(c) (r - 1) (t - 1) (d) None

Q.26 As illustrated in Question 20, the d.f for sampling error is equal to

(a) rt (s -1) (b) rs (t - 1)

(c) ts (r - 1) (d) None

Q.27 As illustrated in Question 20, to test the significance of block and treatment effect, the error M.S. to define the appropriate F statistic will be the

(a) Experimental error M.S
(b) Sampling error M.S
(c) Pooled error M.S
(d) Pooled error M.S. if pooling is not allowed

Q.28 To test the significance of experimental error eij, the test statistic F will follow the distribution given as

(a) F (r - 1) (t - 1), rs (t - 1)
(b) F (r - 1) (t - 1), rt (s - 1)
(c) F (r - 1) (s - 1), rt (s - 1)
(d) F (r - 1) (s - 1), rs (t - 1)

Q.29 If the homogeneity of error variances stands rejected then ANOVA technique can be applied on

(a) Original data
(b) On appropriate transformed data
(c) Changing origin of data
(d) After changing scale of data

Q.30 The least significant test or critical difference test is the

(a) Liberal test as compared to ANOVA
(b) Not liberal test as compared to ANOVA
(c) More complicated test as compared to ANOVA
(d) Slightly complicated test as compared to ANOVA

Q.31 The L.S.D test or C.D. test slightly declares significant

(a) More number of pairs of treatments
(b) Less number of pairs of treatments
(c) Actual number of pairs of treatments
(d) Cannot be stated with surity

Q.32 Duncan's multiple range test is more popular in

(a) Medical Science
(b) Mathematical Science
(c) Agricultural Science
(d) In all the three

Q.33 Duncan's multiple range test compares the
- (a) The range of any set of p means
- (b) The variance of any set of p means
- (c) The C. V.'s of any set of p data
- (d) The standard deviations of p set of data

Q.34 In applying the Duncan's multiple range test, the value of r_p depends upon
- (a) Level of significance
- (b) Error d.f. in ANOVA
- (c) Power of the test
- (d) Both (a) and (b)

Q.35 While applying Duncan's multiple range test, the value of r_p may be obtained from a table known as-
- (a) Critical value of Duncan's multiple range test.
- (b) Table of Critical value of student's t statistics.
- (c) Table of Critical value of variance ratio F statistic
- (d) Table of Critical value of Chi-square statistic

17.4 Key Answers

17.4.1 True / False

1.	True	2.	False	3.	False
4.	True	5.	False	6.	True
7.	True	8.	True	9.	True
10.	True	11.	True	12.	True
13.	True	14.	True	15.	True
16.	True	17.	Frue	18.	True
19.	True	20.	True	21.	True
22.	True	23.	True	24.	True
25.	True	26.	True	27.	True
28.	True	29.	True	30.	True
31.	True	32.	True	33.	True
34.	True	35.	True	36.	True
37.	True	38.	True		

17.4.2 Fill in The Blanks

1. Effect model, Treatments
2. Randome effect model
3. μ, r_i
4. (μ, t_i)
5. Covariance
6. Estimated value, minimum
7. Variablility
8. Treatments
9. n_o, σ_r^2
10. $\sigma^2 + r\ \sigma_r^2$
11. σ^2
12. t_i, e_{ij}
13. Normal,
14. σ^2
15. $s\,\sigma_e^2, +rs$
16. $\sigma^2 + s\sigma_e^2$
17. σ^2
18. $\sigma^2 + s\sigma_e^2$
19. $\sigma^2 + s\sigma_e^2, (s-1)$
20. $\sigma^2, (s-1)$
21. t_i, β_j, e_{ij}
22. s, ts
23. s, ts
24. σe^2
25. σ^2
26. rt
27. (r -1)(t - s)

28. Experimental error M.S
29. (r -1) (t -1) tr
30. Invalid
31. Transformed
32. Liberal
33. More number
34. Rigid, popular
35. Means, $s_{\bar{x}}$
36. Significance, d.f
37. Duncan's multiple

17.4.3 Multiple Choice Questions and Answers

1.	d	2.	c	3.	d
4.	b	5.	a	6.	b
7.	d	8.	a	9.	b
10.	b	11.	b	12.	b
13.	c	14.	d	15.	d
16.	a	17.	c	18.	a
19.	d	20.	d	21.	a
22.	a	23.	c	24.	d
25.	c	26.	a	27.	a
28.	b	29.	b	30.	a
31.	a	32.	c	33.	a
34.	d	35.	a		

18

Transformation of Data to Apply ANOVA Technique

18.1 True / False

Q.1 In addition to the assumption of ANOVA, the other point which is given due consideration by the statistician is "The stability of variance."

Q.2 In any investigation, if it is possible to incorporate the "randomisation procedure" then it is customary to rely on " Randomisation" to break up any "correlation among experimental errors".

Q.3 Two broad categories of heterosedasticity of an experimental error are the "Regular and Irregular type of hetrosedasticity"

Q.4 Irregular type of heterogeneity in errors can be characterised by certain treatment which posses more variability than others.

Q.5 The differences in variabilty of errors in treatments may or may not be expected in advance.

Q.6 To see the effect of an insecticide on number of insects, the number of insects are likely to be more variable in "an untreated experimental units than the treated experimental units".

Q.7 An untreated experimental units contribute more than treated units to the experimental error S.S. or Error M.S.

Q.8 The standard deviation based on "pooled error M.S" will be too large, and it may fail to detect the real differences in effects of various insecticides under study.

Q.9 When the heterogeneity in error is of irregular nature then the best procedure is to omit certain portion of data from the anlysis.

Q.10 If non normality exists in the data then "regular type of heterogeneity arises in experimental error.

Q.11 The variability within several treatments is related to the treatment means in some reasonable fashion.

Q.12 When data obtained from an experiment, deviate from normality then one must apply some suitable transformation to follow the normal distribution.

Q.13 The means and variances of transformed data become independent.

Q.14 The resulting error variances of transformed data, may become homogeneous (stable).

Q.15 When it is not possible to achieve the independence of mean and variance and the homogeneity of error variances through suitable transformation then one should apply "Weighted Analysis Method.

Q.16 If the data follows poisson distribution (i.e mean = variance) then suitable transformation which makes the data suitable to apply ANOVA technique is "Square Root Transformation".

Q.17 In case of data on "Number of Plants" or "Number of insects of a given species" in a given area, one must apply "square root transformation".

Q.18 In case of data on "Number of bacterial colonies in plate"or "Number of bacteria counted in a plate", one should apply square root transformation.

Q.19 For the % data based on counts and with a common denominator where the range of % is within (0 to 20) or (80 - to 100) but not both then suitable transformation is "square root".

Q.20 If the % of count data lies between (20-80) then the % is substracted from 100 before applying the square root transformation.

Q.21 When % values of y happen as y <10 or even when y < 15 and specially when y = 0.00 the transformation is utilised as $\sqrt{y+\frac{1}{2}}$.

Q.22 When the variance (σ^2) of observations taken on treatments is α to (Treatment)^2 or σ α (Treatment) then suitable transformation is the " Logrithmic or log y_{10} before applying ANOVA.

Q.23 If treatment effects are multiplicative on original scale of measurements or when "One Treatment" gives a response consistently 20% higher than others then suitable transformation is "logarithmic one."

Q.24 When the data based on binomial distribution is expressed as decimal fraction or % and % covers a wide range of values then the suitable transformation is the "angular or inversion or arc $\sin\sqrt{y}$ " before applying ANOVA.

Q.25 When $\sin^{-1}\sqrt{y}$ value are expressed in degree, then the variance of such transformed data will be approximately constants, and will be equal to $\left(\frac{821}{n}\right)$ where n is the total number of value in data.

Q.26 When $\sin^{-1}\sqrt{y}$ values are expressed in degree, then the variance of such transformed data is equal to $\left(\frac{1}{4n}\right)$

18.2 Fill in the Blanks

Q.1 In addition to the assumptions of ANOVA, the other point given due consideration by the statistician is the _________

Q.2 It is customary to rely on _________ to break up any "correlation among experimental errors.

Q.3 Two broad categories of heterogeneity of experimental errors are the " _________ and _________" type of heterogeneity.

Q.4 Irregular type of heterogeneity of an experimental error can be categorised by certain treatments which possess more "_________" than others.

Q.5 The number of insects are likely to be more variable in an "_________" than treated experimental units by an insecticide.

Q.6 An untreated experimental units contributes more to the "_________" S.S. or Error M.S. than treated experimental units.

Q.7 The pooled error M.S. being too large, may fail to detect the " _________" effects of various insecticides under study.

Q.8 When the heterogeneity in experimental error is of "_________"nature then the best procedure is to omit certain portion of data from _________

Q.9 When " _________" exists in the data then regular type of heterogeneity arises in "_________".

Q.10 The variability within several treatments is related to the "_________" in same reasonable fashion.

Q.11 When experimental data deviate from normality then one must apply some suitable "_________ before applying ANOVA technique.

Q.12 A suitable transformation applied to experimental data which is non-normal, transforms it to follow approximately the "_________" distribution.

Q.13 The means and variance of transformed data become "_________"

Q.14 The resulting error variance of transformed data may become "_________" in nature.

Q.15 When for the transformed data, it is not possible to make its means and variance "_________" then one should apply "weighted analysis" method.

Q.16 When it does not become possible to achieve the homogeneity of error variances even after transforming the data, then one should apply "_________"

Q.17 If the experimental data follow poisson distribution, then before applying ANOVA technique one should transformed the original data through "_________" transformation.

Q.18 If the experimental data are available on "number of bacterial colonies in plate" or "number of bacteria counted in a plate" then one should apply _________" transformation before applying ANOVA technique.

Q.19 If the experimental data is based on "Number of Plants" or "Number of insects of a given spicy" in a given area, then we should apply "_________" transformation before applying ANOVA technique.

Q.20 If the % data based on counts with common denominator lies in the range (0 - 20) or (80 - 100) but not both then before using ANOVA technique, one should apply "_________" transformation.

Q.21 If the % of count data lies in the range (20 - 80), then % should be "_________" from 100 before applying "_________" technique.

Q.22 When % value, say y happens to be as y < 10 or even y < 15 and specially when y = 0.00, then transformation is expressed as $\sqrt{(....)}$

Q.23 When variance (σ^2) of observations taken on treatments is proportional to (treatment mean)^2 or s.d. (σ) is proportional to (treatment means) then before applying (_________) technique, one should apply the "_________" transformation.

Q.24 If the treatment effects are of multiplicative nature on original scale of measurement then before applying on ANOVA technique, one must apply the " _________" transformation.

Q.25 When out of several treatments, one treatment gives a response consistently 20% higher than others, then before applying an ANOVA technique, suitable transformation applicable is "_________".

Q.26 When an experimental data follow binomial distribution which is expressed as decimal fraction or % and it covers a wide range should apply a suitable transformation known as "_________"

Q.27 When $\sin^{-1}\sqrt{y}$ values are expressed in "_________" then variance of such transformed data becomes approximately constant and is equal to (821/n).

Q.28 When $\sin^{-1}\sqrt{y}$ values are expressed in "_________" then "_________"of such transformed data is equal to 1/4n.

18.3 Multiple Choice Questions and Answers

Q.1 In addition to the assumption of ANOVA, the other point given due consideration by statisticians is the-

(a) Stability of variances
(b) Stability of standard deviations
(c) Stability of treatment means
(d) Both (a) and (b)

Q.2 To break up any correlation among experimental error, its customary to rely on principle of

(a) Replication (b) Randomisation
(c) Local control (d) Both (a) and (c)

Q.3 Irregular type of heterogeneity of an experimental error can be characterised by certain treatment which has more "_________" than others

(a) Variability (b) Variance
(c) Standard deviation (d) All the three

Q.4 An untreated experimental units contribute more to the

(a) Experimental error variance
(b) Experimental error M.S.
(c) Reduction of precision of experiment
(d) All the three

Q.5 The pooled error M.S. being too large, may fail to detect the real differences in effects of various.

(a) Treatments (b) Insecticides
(c) Fertilizers (d) All the three

Q.6 When the heterogeneity in experimental error is of irregular type then the best way to analysis the data is

(a) To omit certain portion of data from analysis
(b) To include some observation in data
(c) To apply the weighted analysis method
(d) Both (b) and (c) jointly

Q.7 If non normality exists in data then "Regular type of heterogeneity".

(a) Dose not arise in experimental error
(b) Arises in experimental error
(c) Does not create problem in analysis
(d) Both (a) and (c)

Q.8 When data obtained from an experiment deviate from normality then one must apply some suitable transformation to follow the

(a) Normal distribution
(b) Poisson distribution
(b) Additive law for treatment and error effects
(d) Homogeneity of error variances

Q.9 Means and variances of transformed data become

(a) Independent (b) Uncorrelated
(c) Correlated (d) Both (a) and (b)

Q.10 When for transformed data, it is not possible to make its mean and variance independent, then one should apply.

(a) ANOVA technique (b) Weighted Analysis
(c) Regression Analysis (d) ANOVA technique

Q.11 When it is not possible to achive the homogeneity of error variances even after applying transformation, one should apply

(a) ANOVA technique (b) Regression Analysis
(c) Weighted Analysis (d) Non parametric method

Q.12 When an experimental data follow poisson distribution, then before applying ANOVA technique, one should transform data through

(a) Logarithmic transformation (b) Square root transformation
(c) Angular transformation (d) Inverse sin transformation

Q.13 If an experimental data are available on "number of bacterial colonies in plate" or "number of bacteria counted in plate" then should apply square root transformation before applying

(a) ANCOVA technique (b) Regression analysis
(c) ANOVA technique (d) Both (b) and (c)

Q.14 If the % data based on counts with common denominator lie in the limit (0 - 20) or (80 - 100) but not in both then one should transform data through square root transformation before applying

(a) ANOVA (b) ANCOVA
(c) Regression analysis (d) Large sample test

Q.15 If the % of count data lie in the range (20 - 80), then % should be substracted from 100 before applying

(a) Square root transformation (b) ANOVA
(c) ANCOVA (d) None of these

Q.16 When % value, say y happens to be as $y < 10$ or even $y < 15$ and specially when $y = 0.00$ then transformation is expressed as

(a) $\sqrt{y+0.5}$ (b) $\sqrt{y+1.0}$
(c) $\sqrt{y^2+0.5}$ (d) $\sqrt{y^2+1.0}$

Q.17 When variance of observation taken on treatment is proportional to $(\text{Treatment mean})^2$ or s.d. is proportional to (Treatment mean) then one should apply logarithmic transformation before using

(a) Regression analysis (b) ANCOVA
(c) ANOVA (d) Both (a) and (c)

Q.18 If treatment effects are of multiplicative nature in original data then one must apply the "logarithmic" transformation before applying

(a) Regression analysis (b) ANCOVA
(c) ANOVA (d) Both (a) and (c)

Q.19 Out of several treatments, if one treatment gives a response consistently 20% higher than others, then before applying ANOVA technique, the suitable transformation is

(a) Logarithmic (b) Square root
(b) Angular (d) Linear

Q.20 When an experimental data follow binomial distribution, which is expressed in % and it covers a vide range of value then before applying ANOVA, the suitable transformation applied is

(a) Square root (b) Angular
(c) Logarithmic (d) None

Q.21 When $\sin^{-1}\sqrt{y}$ values are expressed in "degree" then the variance of such transformed data is approximately equal to

(a) (621/n) (b) (721/n)
(c) (821/n) (b) (521/n)

Q.22 When $\sin^{-1}\sqrt{y}$ values are expressed in "radian" then variance of that transformed data is approximately equal to

(a) (1/4n) (b) (1/2n)
(c) (2/8n) (b) Both (a) and (c)

18.4 Key Answers

18.4.1 True / False

1.	True	2.	True	3.	True
4.	True	5.	True	6.	True
7.	True	8.	True	9.	True
10.	True	11.	True	12.	True
13.	True	14.	True	15.	True
16.	True	17.	True	18.	True
19.	True	20.	True	21.	True
22.	True	23.	True	24.	True
25.	True	26.	True		

18.4.2 Fill in the Blanks

1. Stability of variances
2. Randomisation
3. Regular, Irregular
4. Variabitily
5. Untreated experimental unit
6. Experimental error
7. Real differences
8. Irregular, Analysis
9. Non normality, Experimental error
10. Treatment means
11. Transformation
12. Normal
13. Independent
14. Homogeneous
15. Independent
16. Weighted Analysis
17. Square root
18. Square root
19. Square root
20. Square root
21. Substracted, ANOVA
22. $\left(y+\frac{1}{2}\right)$
23. ANOVA, Logarithmic
24. Logarithmic
25. Logarithmic
26. Angular or $\sin^{-1}$
27. Degree
28. Radian

18.4.3 Multiple Choice Questions and Answers

1.	d	2.	b	3.	d
4.	d	5.	d	6.	a
7.	b	8.	a	9.	d
10.	b	11.	c	12.	b
13.	c	14.	a	15.	a
16.	a	17.	c	18.	c
19.	a	20.	b	21.	c
22.	d				

19

Non Parametric Methods

19.1 True / False

Q.1 The most fundamental assumption to be followed by random variable for using the standard method (parametric statistic) or parametric inference is that the random variable follows normal distribution.

Q.2 The standard parametric techniques are not useful for estimating parameters and testing hypothesis about them.

Q.3 Any sample statistic which is used to estimate a parameter and to test a hypothesis concerning it for a population is known as parametric statistic.

Q.4 For a considerable amount of data collected, if the underlying distribution is not specified then the "distribution free statistic" (non parametric method) is appropriate to handle that data.

Q.5 If the sampling distribution of a sample statistic is free from the parameters of the parent distribution from which the sample statistic has been derived, is not known as "distribution free statistic or non-parametric statistic".

Q.6 The sampling distribution of a non-parametric statistic never depends upon the parameters of the parent population.

Q.7 The non parametric statistic compares the distributions rather than parameters where as the parametric statistic compares the parameters of the distribution.

Q.8 The non parametric statistics may be sensitive to "change in location or spread or both characteristics" of the population

Q.9 When only weak assumptions are made about the distributions for underlying data, then non parametirc statistic is more suitable than parametric statistic for that data.

Q.10 When there is possiblitly to rank the available data, then non- parametric procedures are not available to handle such data.

Q.11 If the observations are in the form of continuous, or in ranks, or in the signs of differences of paired observations then non parametric statistics are appropriate.

Q.12 If the exact form of distribution function of a parent population is known or if the data can be transformed in such a way that the transformed data follow some specified distribution, then N.P procedure extracts less information than that which is really available in data.

Q.13 If all the experiments of an investigation result in the data such that the Ho is true then in that situation parametric procedure is not as good as non-parametric procedure.

Q.14 If Ho is false and assumptions about parent population are valid then "Parametric procedures" (ie classical procedures) are found better, than non-parametric procedures.

Q.15 In case of small sample sizes, say $n \leq 10$, the efficiency of non-parametric procedure is quite high as compared to that to parametric procedure.

Q.16 The sign test is based on the sign of differences obtained between the paired values and the test statistic used is X^2 defined as-
$X^2_1 = (n_1 - n_2)^2 / (n_1 + n_2)$
Where (n_1 & n_2) are the number of (+ & - 1) signs.

Q.17 To maintain the continuity property of X^2 distribution, in case of applying sign test, as in Q.16, one should consider the numerator $(|n_1 - n_2| - 1)^2$ in peace of $(n_1 - n_2)^2$ as an adjustment made.

Q.18 In case of applying sign test, only the numbers of (+ & -) sign of differences (Xi - Yi), of paired observations (Xi , Yi, i = 1, 2, ...n) are "demerit" of sign test.

Q.19 In case of number of observations being fewer than six, it is always possible to detect the departure from Ho hypothesis.

Q.20 If one has to test Ho for a population, that the median is a specified value say Md then one must use the "sign test" with test statistic X^2_1 defined as $X_1^2 = \dfrac{(n_1 - n_2)^2}{(n_1 + n_2)}$

Q.21 A signed rank test is considered as an improvement on sign test because it uses the magnitude of differences also.

Q.22 If (T^+ and T^-) are the sum of (+ve and -ve) ranks in case of adopting signed rank test respectively then numerically lower value of (T^+ , T^-) will be considered as "test statistic".

Q.23 In case of signed rank test, if "test statistic" < "tabulated value" then Ho is declared significant while in case of other tests (parametric) like, Z, F, t and X^2, if "test statistic" < tabulated value then Ho is declared non-significant.

Q.24 If the number of paired observation is beyond range of table (Say n > 25), then one should adopt Z test based on fundamental area property.

Q.25 Let in case of applying signed rank test, the value of "test statistic" be T and the number of paired observations say "n" > 25, then z is defined as-

$$Z = (T - \mu_T) / \sigma_T, \text{ where } \mu_T = \frac{n\,(n+1)}{4}, \sigma_T = \sqrt{\frac{n(n+1)(2n+1)}{24}}$$

Q.26 In case of "Wilcoxon Mann Whitney two sample test" if the test statistic (say %) < tabulated value then Ho will be declared significant.

19.2 Fill in the Blanks

Q.1 The most fundamental assumption for parametric statistic or parametric inference is that the random variable follows ________________ distribution.

Q.2 The standard parametric methods are useful to ________________ population parameters and to test the ________________ about them.

Q.3 Any sample statistic used to estimate the populations parameter and to test a null hypothesis about it is known as ________________

Q.4 For a considerable amout of data collected, if the underlying distribution is not specified then the appropriate method to handle such data is the ________________ or ________________

Q.5 If the sampling distribution of a sample statistic is independent of parameters of parent population then it is known as "________________" statistic or "________________" statistic.

Q.6 The sampling distribution of a non-parametric statistic is always "________________" of the parent population parameters.

Q.7 The non parametric statistic compares the ________________ rather than parameters of the distributions.

Q.8 The parametric statistic compare the ________________ rather the distribution.

Q.9 When only weak asumptions are made about the distributions for underlying data then "________________" statistic is suitable than "________________" statistic for that data.

Q.10 When there is possibility to rank the available data, then "_______________" procedures are suitable to handle such data.

Q.11 When observations are in continuous form or in ranks or in signs (+ & -) of difference of paired values then "_______________" statistics are appropriate to be applied.

Q.12 when the exact form of distributions function of a parent population is known the N.P. procedure extract "_______________" information than that which is realy contained in data.

Q.13 When the transformed data of original one follow some specified distribution then appropriate statistic for the transformed data is the "_______________" statistic.

Q.14 If all the experimentas of an investigation result in such data that Ho is true then "_______________" procedure as good as non-parametric procedure.

Q.15 If the assumptions about parent population are valid but Ho is false then "_______________" procedure are found better than non parametric procedures.

Q.16 In case of small sample sizes, say n < 10, the efficiency of "_______________" procedure is as quite high as compared to that of parametric procdure.

Q.17 For the signed test applied to paired values of a sample the test statistic X_1^2 is defined as.

$$X^2 = (\quad)^2 / (\quad)$$

where (n_1, n_2)are the numbers of (+, - 1) signs.

Q.18 In Question 17, to maintain the continuity property of X^2 distribution, the numerater is replaced by (_______________)^2 as an adjusted value.

Q.19 In case of applying "_______________" only numbers of (+, -) sign of differences (Xi -Yi) are considered to define test statistic.

Q.20 If number of observation $\angle 6$, it is not possible to detect the departure from "_______________"

Q.21 For testing the null hypothesis:

Ho: Median of a population is equal to Md then for "_______________" the test statistic is "_______________" which is defind as.

$$X_1^2 = (n_1 - n_2)^2 / (n_1 + n_2)$$ in usual notations.

Q.22 A "__________" is considered as an improvement on "__________" because it consider the magnitude of the difference (Xi - Yi).

Q.23 Let (T^+, T^-) be the sum of (+ ie, - ve) ranks in case of adopting "__________" then numericaly lower value of (T^+, T^-) will be considered as "__________" .

Q.24 In case of applying "__________", if -

test statistic < tabulated value, then Ho is declared "__________"

Q.25 If number of paired observation is beyond range (ie n>25) then one should apply the "__________" based on fundamental area property of normal distribution.

Q.26 To apply the signed rank test if the value of "__________" is T and number of paired observation say "n" > 25, then Z is defined as-

$Z = (T - \mu_T)\sigma_T$ where $\mu_T =$. __________ , and $\sigma_T = \sqrt{\ldots\ldots\ldots}$

Q.27 In case of applying "Wilcoxon Mann Whitney" two sample test, if- test statistic (say T) < tabulated value then Ho is declared__________"

19.3 Multiple Choice Questions and Answers

Q.1 The most fundamental assumption for parametric inference is that the random follows

(a) Continuous distribution (b) Normal distribution
(c) Non-normal distribution (d) Student's distribution

Q.2 The standard parametric methods are useful to estimate the

(a) Population parameters (b) Sample statistic
(c) Population constants (d) Both (a) and (c)

Q.3 Any sample statistic which is used to estimate some population parameter and to test hypothesis about it is known as "__________" statistic.

(a) Non - Parametric (b) Parametric
(c) Estimator (d) Test

Q.4 If the underlying distribution for a considerable amount of data collected, is not specified one, then the suitable method to deal with data is known as

(a) Parametric method (b) Non - Parametric method
(c) Distribution free method (d) Both (b) and (c)

Q.5 If the sampling distribution of a sample statistic is independent of parent population parameter then that statistic is known as

(a) Distribution free statistic (b) Parametric statistic
(c) Non-parametric statistic (d) Both (a) and (c)

Q.6 The sampling distribution of a non - parametric statistic is always-

(a) Free from parent population parameter
(b) Independent of parent population parameter
(c) Dependent of parent population parameter
(d) Both (a) and (b)

Q.7 Non parametric statistic (method) compares the

(a) Distribution (b) Parameters
(c) Estimates (d) Test statistics

Q.8 Parametric statistic (Method), compares the

(a) Parameters of the distribution
(b) Estimates of the population parameters
(c) Estimates of the population constants
(d) All (a), (b) and (c)

Q.9 When only weak assumptions are made about the distributions for underlying data, then an appropriate method to handle such data is the

(a) Parametric method (b) Non-parametric method
(c) Distribution free method (d) Both (b) and (c) method

Q.10 When there is a possibility to rank the available data, then suitable method to analyse that data is

(a) Parametric method (b) Non-parametric method
(c) Distribution free method (d) Both (b) and (c) method

Q.11 When observations are in continuous data form or in rank form or in sign (+ 8 -) of differences ($x_i - y_i$) of paired values ($x_i - y_i$) the appropriate method to analyse such that is

(a) Non-parametric method (b) Parametric method
(c) Distribution free method (d) Both (a) and (c)

Q.12 When the exact form of distribution functions of a parent population is known then "Less information" is extracted from the data if we apply the

(a) Non - prametric method (b) Parametric method

(c) Distribution free method (d) Both (a) and (c)

Q.13 When transformed data of original one, follow some specified distribution then appropriate statistic for such transformed data is

(a) Parametric (b) Non-parametric

(c) Distribution free (d) Both (b) and (c)

Q.14 If the assumptions about the parent population are valid but Ho is false then appropriate procedure to deal with such data is

(a) Parametric procedure (b) Non parametric procedure

(c) Distribution free method (d) Both (b) and (c)

Q.15 To apply the signed test to the paired values (xi, yi) of a sample, the test statistic X^2 is defined as

(a) $X^2 = \frac{(n_1 - n_2)}{(n_1 + n_2)}$ (b) $X^2 = \frac{(n_1 - n_2)^2}{(n_1 + n_2)}$

(c) $X^2 = \frac{(n_1 + n_2)}{(n_1 - n_2)}$ (d) $X^2 = \frac{(n_1 + n_2)}{(n_1 - n_2)^2}$

Q.16 In Q.15, to maintain the continuity property to X^2 distribution, the numerator is replaced by

(a) $(|n_i - n_2|+1)^2$ (b) $(|n_i - n_2|-1)^2$

(c) $(|n_i + n_2|-1)^2$ (d) $(|ni - n2|-1)^3$

Q.17 Only numbers of (+, - 1) sign of differences (X_i - Yi) are considered to define the test statistic if we apply the test known as

(a) Sign test (b) Signed rank test

(c) Chisquare test (d) Run test

Q.18 For testing the null hypothesis
Ho : Median of a population is equal to Md.

the sign test is applied with X_1^2 defined as

(a) $X^2 = \left(\frac{n_1 - n_2}{n_1 + n_2}\right)$ (b) $X^2 = \left(\frac{n_1 + n_2}{n_1 - n_2}\right)$

(c) $X^2 = \frac{(n_1 - n_2)^2}{(n_1 + n_2)}$ (d) $X^2 = \frac{(n_1 + n_2)}{(n_1 - n_2)^2}$

Q.19 A signed rank test is considered as an improvement of

(a) Run test (b) Sign test

(b) Median test (d) Non-parametric test

Q.20 Let (T^+, T^-) be the sum of (+ ve, - ve) ranks in case of adopting signed rank test then numerically lower value of (T^+, T^-) will be considered as

(a) Test statistic

(b) An estimate

(c) An estimator

(d) Lower limit of interval estimate

Q.21 In case of adopting signed rank test, if
Test statistic < Tabulated value then Ho is

(a) Declared non significant (b) Declared significant

(c) Not decidable (b) None

Q.22 If the number of paired observations say n > 25 then based on fundamental area property of normal distribution, one must apply the

(a) Z test (b) Non-normal test

(c) F test (b) X^2 test

Q.23 To apply the signed rank test if the value of test statistic is T and number of paired observations say n > 25 then z is defined as

(a) $Z = \frac{|(T - \mu_T)|}{\sigma_T}$ where $\mu_T = \frac{n(n+1)}{2}$

(b) $Z = \frac{|\mu_T - T|}{\sigma_T^{\,2}}$ where $\mu_T = \frac{n(n+1)}{4}$

(c) $Z = \frac{(T - \mu_T)}{\sigma_T}$ where $\mu_T = \frac{n(n+1)}{2^2}$

(d) Both (b) and (c)

Q.24 In case of applying "Wilcoxon Mann Whitney" two sample test, if Test statistic (say T) K< Tabulated value then Ho is declared

(a) Significant (b) Non-significant

(c) Rejected (d) Both (a) and (c)

19.4 Key Answers

19.4.1 True / False

1.	True	2.	False	3.	True
4.	True	5.	False	6.	True
7.	True	8.	True	9.	True
10.	False	11.	True	12.	True
13.	False	14.	True	15.	True
16.	True	17.	True	18.	True
19.	False	20.	True	21.	True
22.	True	23.	True	24.	True
25.	True	26.	True		

19.4.2 Fill in The Blanks

1. Normal
2. Estimate, hypothesis
3. Parametric statistic
4. Parametric statistic or Distribution Free method
5. Non parametric or Distribution free
6. Independent
7. Distributions
8. Parameters
9. Non-parametric, Parametric
10. Non-parametric
11. Non-parametric
12. Less

13. Parametric
14. Parametric
15. Parametric
16. Non-parametric
17. $(n_1 - n_2)^2 / (n_1 + n_2)$
18. $(|n_1 - n_2| - 1)$
19. Sign Test
20. Null hypothesis
21. Sign Test, Chisquare
22. Signed Rank Test, Sign Test
23. Signed Rank Test, Test Statistic
24. Signed Rank Test, Significant
25. Normal test
26. Test Statistic, $\frac{n(n+1)}{4}$

$$\sqrt{\frac{n(n+1)(2n+1)}{24}}$$

27. Significant

19.4.3 Multiple Choice Questions and Answers

1.	b	2.	d	3.	b
4.	d	5.	d	6.	d
7.	a	8.	d	9.	d
10.	d	11.	d	12.	d
13.	a	14.	a	15.	b
16.	b	17.	a	18.	c
19.	b	20.	a	21.	b
22.	a	23.	d	24.	d

20

Additional Topics on Sample Surveys

SECTION-A : TRUE/FALSE

20.1 Sampling on Successive Occasions

Q.1 In applying varying probability sampling for information of population total of X, it is required to know the values of selection probabilities Pi (i = 1, 2, 3,..., N).

Q.2 The technique of "Double Sampling" was first time formulated by Jayman, (1938) in connection of collecting information on strata sizes in stratified random sampling scheme.

Q.3 When the sample for main survey is selected in three or more phases, the procedure is termed as multiphase sampling.

Q.4 In sociological and economic research, one is offen in need of measuring the characteristic of a population on several occasions to estimate the population means as a time series/current value of population mean.

Q.5 When the same population is sampled repeatedly, the opportunities for flexible sampling designs are greatly enhanced.

Q.6 On hth Occassion, we may have a part of sample matched with the sample at $(h\text{-}l)^{th}$ occassion. Such parts may be matched with both $(h\text{-}l)^{th}$ and $(h\text{-}2)^{th}$ occassion. etc. Such method of partial matching, is termed as "Sampling on Successive Occassions with partial replacement of units (Patterson 1950).

Q.7 The sampling on two occasions was pioneered by Jessen (1942) for estimating the current population mean.

Q.8 When there are more than two occassions, one has more flexibility in using both the sampling schemes and the estimates of the characteristics under study.

Q.9 The loss of efficiency occurred by using the information from the latest two or three occasions only, is fairly small as compared from many occassions.

Q.10 If the probability of happening of a rare characteristic is very small then method of "Inverse sampling" is quite applicable.

Q.11 In case of inverse sampling scheme, the sample size n is not fixed in advance,

Q.12 In inverse sampling scheme, the sampling is continued until a predetermined number of units possessing the rare attribute is included in the sample.

Q.13 Inverse sampleing scheme is also known as "Sequential sampling scheme."

20.2 Non-Sampling Errors

Q.14 In complex surveys, it is not possible to get true values on all the attributes of interest on all units in the sample.

Q.15 Non-response may occur due to ommission/laps on part of an interviewer or refusal on part of respondent.

Q.16 Non-response may not occur due to non-availability of an individual (respondent) during the period of survey.

Q.17 Non-response may occur when respondents refuse to reply some specific questions.

Q.18 Non-response may not occur due to skiping some questions by the respondents/interviewer.

Q.19 An "Error in measurement" may occur due to biased or inaccurate measuring device.

Q.20 Non-response error occur in the phase of data collection.

Q.21 An increased interviewer-training, repeated call backs, sub-sampling of non-respondents, simplification of questionnaire and randomised response technique, may minimise non-response error in a survey.

Q.22 In sample survey, an incentive for co-operation can be used to maximise response rates.

Q.23 In case of adjustment for unit non-response, in general, two types of substitution procedures are used.

Q.24 In case of non-response, two types of substitution procedures are :

i) Selection of a random substitue and -

ii) Selection of a specially designated substitute.

Q.25 In a random substitution procedure, an additonal unit is selected to replace each non-responding unit generally from the same area or stratum to which non-responding units belong.

Q.26 In specially designated substitute, one or more backup units are identified to replace each sampled unit beforoe start of the surveys.

Q.27 A common procedure in dealing with non-response in surveys is to call back the unit, before a preassigned number of times completely giving up the unit as a non-respondent.

Q.28 The call back estimators $\overline{y_i}$ $(i = 1, 2, 3, ...k)$ are biased if the populations in different response classes have different means.

Q.29 In sample surveys, it may happen that the informations for some items of the questionnaire are provided but not for other items. The plausible values for missing items are inputed while processing such records.

Q.30 Imputation is the process of explicity/implicity substituting plausible values for incomplete/inconsistent items in survey records.

Q.31 Quata sampling is a process which enables one to obtain a desired number of completed cases in a relatively short period of field work without expense of call backs.

Q.32 In traditional sampling procedures, socio-economic surveys often give inefficient results because of missing data arised due to undercoverage and non-response errors.

Q.33 To tackle the problem of non-response in sample surveys, Warnor (1965) introduced a "Randomised response technique (R.R.T.) which has been develped subsequently by variaus authors.

20.3 National Sample Surveys

Q.34 The national sample survey (NSS) was initiated in 1951 for conducting sample surveys to provide socio-economic data to government and other agencies.

Q.35 The data collected by NSS can be utilized to make plans for National development and various research purposes.

Q.36 NSS conducts continuing surveys which are carried out in the form of round.

Q.37 Each round of NSS covers some important topics of current interests.

Q.38 Duration of each round of NSS varies (from 3 to 8) months.

Q.39 NSS in its first few rounds, gave more emphasis to collect information for computing the national income.

Q.40 To work out the national income, the NSS collected statistics (data) for expenditure and small scale house hold enterprises.

Q.41 In 8th round, the NSS shifted to study the distribution of land holdings.

Q.42 In 9th round, the primary objective of NSS was to study the employment and unemployment situation in India.

Q.43 In 10th round of NSS, a study was made related to the estimation of acrage and yield rate of serial crops alongwith socio-economic enterprise.

Q.44 In 11th and 12th round of NSS, a study related to economic conditions of agriculture in India was made.

Q.45 In 13th round of NSS, study was made on crop surveys.

Q.46 Statistical information about a place, region, state or nation in cecessary to run the administration efficiently.

Q.47 The land revenue rate is based on agricultural production and fertility level of soil.

Q.48 The first census in India covering whole of the country was conducted in 1881.

Q.49 In order to collect and compile the economic data at state level, a statistical buereau was established at centre in 1895.

Q.50 An enquiry committtee known as "Economic Enquirey Committee" was established in 1925 under the chairmanship of Sir M. Visvesraya.

Q.51 Immediately after the reporot of "Economic Enquiry Committee" the report of "Royal Commission on Agr. was made out in 1928.

Q.52 In inportent event which occured in 1933 was the establishment of "Statistical Research Bureau."

Q.53 To ensure the completeness of statistical data on industries, an act known as "Industrial Statistics Act" was passed in 1942.

Q.54 Statistical information is the backbone of government activities and planning which has been paid great attention after the independence of India in 1947.

Q.55 The cabinate secretariate has a department of statistics.

Q.56 One wing of department of statistics of cabinate secretariate is the "National Sample Survey Organization (NSSO) Set up in 1950.

Q.57 Central govt. established (C.S.O.) under cabinate secreteriate.

Q.58 The objective of C.S.O. is to create co-ordination of large variety of statistical information collected at center and state level.

Q.59 C.S.O. does the advisory work related to statistical matter, especially the standardization of concepts and definitions to maintain an uniformity throughout the country.

Q.60 C.S.O. collects statistical data related to planning.

Q.61 C.S.O. compiles the national income estimates.

Q.62 C.S.O. provides statistical data of nation to united nations statistical offices and other institutions.

Q.63 C.S.O. attends the work of "International statistical Institute (conferences)" held in India and abroad.

Q.64 C.S.O. provides training to statistical personnel.

Q.65 C.S.O. co-ordinates statistics collected by different ministries and N.S.S.O.

Q.66 C.S.O. supplies information related to methodology adopted in the collection of data, coverage and scope.

Q.67 Directorate of "Industrial Statistics" was set up in 1944 at Kolkatta.

Q.68 C.S.O. lays down the definitions and concepts in conformity with international standard.

Q.69 The following are the international set up.

i) International Standard Industrial Classification (ISIC).

ii) International Standard Classification of Occupation (ISCO).

iii) Standard Occupational Classification (IOC) for India.

Q.70 A planning cell was set up in C.S.O. in 1955 to maintain liaison between ISI and planning division.

Q.71 All statistical matters related to planning are dealt within in "Planning cell" of C.S.O.

Q.72 C.S.O. works out the national income estimates for centre and states in collaboration with various statistical agencies at state level.

Q.73 A National Income Unit (NIU) was created in 1955.

Q.74 The pertinent information on administration and public interest is displayed through graphs, charts and diagrams.

Q.75 C.S.O. collects & compiles economic activities like-motor vehicles, civil aviation, foreign trade, inland trade, taxation and revenue, Agri, forestry, fishery and livestock.

Q.76 The following are the C.S.O.'s publications :

i) Monthly Abstract of Statistics

ii) Monthly Statistics of Production of Selected Industries.

iii) Annual Survey of Industries.

iv) National Accounts Statistics (NAS).

Q.77 NSSO was established in Jan 1950 in the "Department of Economic Affairs" Ministry of Finance.

Q.78 NSSO conducts country wide multipurpose sample surveys covering all aspects of national income for the use of "National Income Committee (NIC)."

Q.78 NSSO conducts sample surveys for planning commission and other ministries of "Govt. of India."

Q.79 Directorate of NSS works under the "Deptt. of State" of cabinate secretariate.

Q.80 The Directorate of N.S.S. does the following works :

1. It collects socio-economic data related to the demographic conditions of the whole country on regular basis.
2. Provides statistical data for national income and planning.
3. Conducts annual survey in the organised industrial sector.
4. It gives training to personnel and provides guidence to states for conducting surveys.

Q.81 NSSO works under the dual control of two sectors.

i) Field work is carried out by the "Directorate of NSS.

ii) Designing of surveys, processing of data and preparation of reports is done through I.S.I.

Q.82 NSSO provides methodology and analyses of information to be utilised by research workers.

Q.83 NSSO informs to public about the new developments in economic and social activities.

Q.84 NSSO has the unified control over the governing council related to survey design, field operation, data processing, economic analysis and publication of NSS data.

Q.85 NSSO consists of four functional divisions with a "chief executive officer" at the apex.

Q.86 Four divisions of NSSO are :

i) Survey design and research.

ii) Field operation

iii) Data processing

iv) Economic analysis.

Q.87 The executive officer of NSSO as the member secretory of "Governing Council" obtains the approval of council to the programme of work and directs the appropriate division to implement it.

Q.88 In concern of Agr., the field operation division (F.O.D.) is responsible for all activities.

Q.89 The FOD with its 41 regions is grouped into 5 zones with H.Q. at Bangalore, Nagpur, Jaipur, Allahabad and kolkatta.

Q.90 Each zone is placed under the charge of a "Deputy Director who is responsible to organise training courses for its zonal staff.

Q.91 Vital statistics is the branch of statistics which deals with the registration of facts related to birth, marriage, divorce, sickness and death under the state direction.

Q.92 The two methods of census are :

1. Defacto method or date system

2. Dejure method or period system

Q.93 First census of free India was conducted in 1951.

Q.94 Census act of India was passed in 1948.

Q.95 The officer in-charge of census office works as "Census commissioner" during the period of census and carries out the duties of "Registrar General" during inter-census period.

Q.96 It was for the first time in 1951 that a distinction between "House" and "House-Hold" was made.

Q.97 House was defined as "Dwelling place with a seperate main gate."

Q.98 House hold was defined on basis of Chulha i.e. "Persons or group of persons living together and dining from a common kitchen were considered as the member of the same household."

Q.99 The questionnare of 1961 census was prepared in consultation with planning commission.

Q.100 The questionnare of 1961 census consisted of two parts namely :

i) Household schedule

ii) The individual slip

Q.101 The C.S.O. has taken a major step in 1977 in collaberation with "State statistical Bureaus" to remove data gaps in the un-organized sectors of non-agricultural economy.

Q.102 The economic census surveys collected information related to the quantity of inputes, output and investment, employment etc.

Q.103 The "Registrar General of India" was responsible to conduct economic census surveys.

Q.104 The 1981 economic census survey is a phenomenon of national importance and is the base for all the future planning and development programmes.

20.4 Agricultural Statistics System in India

Q.105 In ancient days, all the rulers maintained the records of agricultural land and their categories on the basis of fertility and production.

Q.106 Ancient rulers collected money in the form of "Land Revenue".

Q.107 All the statistics which have an impact on agricultural economy of the country may be regarded as "Agricultural Statistics."

Q.108 The statistics related to land utilization, crop production, livestock, agricultural prices, poulty and forestry etc come under the category of "Agricultural Statistics."

Q.109 Agricultural Statistic is mostly collected by the "Directorate of Economics and Statistics (DES)" at centre as well as state level.

Q.110 Each state has a "Directorate of Economics & Statistics."

Q.111 In agricultural statistics, the statistics regading area, crop production, poutty and forests are collected.

Q.112 The total area statistics are maintained from two sources:

i) The surveyor General of India,

ii) The village records maintained by the "Revenue Department."

Q.113 The area under crops is obtained by two sources: i) Offical series based on village records, ii) NSS series based on sample surveys.

Q.114 The NSS collects data during the regular rounds of survey on area under different crops.

Q.115 The land utilisation statistics are published in "Indian Agricultural Statistics".

Q.116 The individual states publish agricultural statistics in "Season and crop Reports."

Q.117 Land utilisation statistics are also published in "

i) Agricultural Situation in India (Monthly)

ii) Abstract of Agricultural Statistics. (DES Annual)

iii) Statistical Abstract of Indian Union (C.S.O. Annual).

Q.118 Agricultural production includes the production of food and non-food crops but forests, livestock and fisheries are excluded at this stage.

Q.119 The output of agricultural production is estimated by multiplying the area under a crop by the average expected yield/hectare., in the season. The yield statistics are collected by official machinary and NSSO.

Q.120 Two methods are adopted to collect yield statistics of various crops, as (i) The traditional method and (ii) The random sampling method.

Q.121 The traditional methods of yield estimation is known as "Annawari system."

Q.122 Under the traditional method of yield estimation, the crop condition is judged by the Patwari or any other official in relation to the normal crop.

Q.123 The crop condition is known as "Condition factor or the Annawari Estimate."

Q.124 In annawari system, we have :

$$\text{Crop yield} = \frac{\text{Annas Judged}}{\text{Total Annas}} \text{ x normal yield x area.}$$

Q.125 The ICAR introduced the "Random Sampling Method" in 1942 for estimating cotton crop.

Q. 126 In random sampling method, village were randomly selected in tahsils of a state and within each village, few fields were selected randomly.

Q.127 Actually for crop yield estimation, the "Stratified Multistage Random Sampling Design." is adopted for the "Crop Cutting Experiments".

Q.128 Now the crop cutting expts. are conducted by the "Directorate of Economics and statistics" and the statistical section of the "Board of Revenue" through its field staff and separately by NSSO.

Q.129 The NSSO conducts crop cutting experiments on major cereal crops. during the course of their regular survey rounds.

Q.130 First a detailed Agriculture Census in India was conducted in the year 1970-71 (from July 1,1970 to 30th June, 1971).

Q.131 During the first Agriculture Census in India, the information was collected about the land holding, land use under different crops (irrigated and unirrigated) and water source-wise irrigated land etc.

Q.132 The second Agriculture Census was conducted during the agricultural year (1976-77) i.e. (from July 1, 1976 to 30th June, 1977).

Q.133 During second Agricultural Census in India, one Kharif and one rabi season were covered completely.

Q.134 Agricultural Census (1980-81) is the 3rd of its kind in India, sponsered by "United Nations Food" and Agricultural Organization (F.A.O.).

Q.135 In India, Agricultural census is conducted under the Ministry of Agriculture" in centre and the "Department of Revenue" in the states.

Q.136 In Agricultural Census, Department of Agriculture, district statistical officer and other government officials participated actively.

Q.137 Another agricultural survey named as "Input Survey" was conducted during the agricultural year (1981-82).

Q.138 During the survey year (1981-82), enumerators collected the information in a single visit for both the seasons.

Q.139 The first cattle census, on all India basis, was conducted (from December, 1919to April, 1920).

Q.140 Now, the livestock census is held quinquennially (At the interval of 5 years) by the "Directorate of Economics and Statistics."

Q.141 The statistics, regarding livestock and poultry are published in"Indian Livestock Census quinquennially.

Q.142 Some livestock popular publications are :

1. Indian Livestock Census. (Quinquennially)
2. Indian Livestock Statistics (Annual)
3. Statistical Abstract of India

Q.143 Forest statistics includes the information relating to the volume of timber, the round wood, the match and pulp wood, the bamboo, the sandal wood, the charcoal and the lac etc.

Q.144 Data relating to area, the volume of standing timber, the annual outturn, the value of timber are released annually in the "Indian Forest Statistics) by D.E.S.

Q.145 For major products statewise, the data on quantities and royalty value are published annually in the "Forestry in India." in a mimeograph form (Duplicating machine) with a time lag of (3 or 4 years).

Q.146 Importent publication on forestry are :

1. Abstract of Agricultural Statistics (D.E.S.)
2. Forestry in India (Annual)
3. Statistical Abstract of India (C.S.O. Annual)
4. Annual Administrative.

Q.147 Fishery statistics is available only in published reports of the "Deptt. of Fisheries" Madras & Bengal.

Q.148 Data about commercial fishing is based on two sectors

i) Ocean coastal and offshore water fishing.

ii) Inland waters e.g. rivers, canals, tanks etc.

Q.149 The estimates of the value of the output of fish are worked out separately for marine & inland fish at the state level.

Q.150 The data on landing of marine fish are collected by the "Central Marine Fisheries Research Institute" C.M.F.R.I., Cochin.

Q.151 The National Income Committee (NIC) was set up by the Govt. of India in 1949.

Q.152 The estimates alongwith the methodology., were published in the first and final report of the NIC. (Ministry of Finance) in (1951 & 1954).

Q.153 The first results of "National Income" estimates were presented by the C.S.O. in "National Income Statistics."

Q.154 The "Sum of the employees compensation and the net income from property and enterpenurship, that is distributed factors income, represents the "National Income of the country."

Q.155 In simple words, "The value of commodities and services produced during a particular period counted without duplication" is the national income.

Q.156 National Income estimate gives us an idea of the purchasing power of the people in the country.

Q.157 National income statistics provides a useful guide, line in the formulation of budget of the country.

Q.158 National income statistics have been found to be useful to make a study the problems of the economically underdeveloped countries.

Q. 159 National income statistics provides a basis for the future planning of a country.

Q.160 The following three methods are used to estimate the national income :

1. Output or production method
2. Income method
3. Expenditure method

Q.161 The output method is also known as "Inventory Method" and it consists in finding out the market value of all goods & services produced by the individuals and business enterprises during a speciafied time-period.

Q.162 National income = (Value of goods & services + self consumption + increase in stock - depreciation of capital - income from abroad).

Q.163 In income method, the national income consists of adding together all incomes by the way of wages, interests, rents & profits.

Q.164 In expenditure method, national income is equal to the sum of consumption and saving i.e. $y = c + s$

where y : National income, c ; consumption and s ; saving.

Q.165 Financial statistics are the best sources to understand the status of a country.

Q.166 R.B.I, has a full fledged department of research and statistics.

Q.167 RBI compiles data and publish report on banking exchange, currency and finance etc.

Q.168 Financial statistics compiled by RBI is used by Govt. of India, Foreign Govt. and International Monetary Fund (IMF).

Q.169 Some important publications of RBI are :

1. RBI Bulletin (monthly)
2. Statistical Tables Related to Banks in India (Annual)
3. Report on Currency and Finance (Annual)
4. Report on Trend and Progress of Banking in India. (Annual)
5. Survey of Finance of Local Authorities (Bulletin)
6. Study of Company Finance (Bulletin)
7. Finance Accounts of Central & State Govt. (Annual)

SECTION-B : FILL IN THE BLANKS

20.1 Sampling on Successive Occassions

Q.1 The techniqe of "Double Sampling" was first time formulated by ________ in (1938).

Q.2 When the sample for main survey is selected in three or more phases, the procedure is termed as "________".

Q.3 When same population is sampled repeatedly, the opportunities for a flexible "________" are enhanced.

Q.4 Sampling on two occassions was first time reported by "________" to estimate the current population mean.

Q.5 In case of more than two occasions, one has a large flelxibility in using both "________" and "________" of the characteristic under study.

Q.6 The loss of efficiency occurred by using information from latest two or three occassions only is fairly "________" in many occassions.

Q.7 If the pobability of happening of a rare attribute is very "________" then method of_______ is applicable successifully.

Q.8 In case of inverse sampling scheme, the sample size n is "________" in advance.

Q.9 Inverse sampling scheme is also termed as "________".

20.2 Non-Sampling Errors

Q.10 Non-response may occure due to non-availability of "________" during the period of survey.

Q.11 Non-response may occure due to "________" or lapse on part of "________" or refusal on part of ''________".

Q.12 An "________" may occure when respondents refuse to answer specific questions.

Q.13 An item non-response may occure because the "________" or the "________" may skip the questions.

Q.14 In case of sensitive questions, the respondents may give "________" intentionally.

Q.15 Due to the measuring device being biased or inaccurate, an "________" may occure.

Q.16 "________" error occure in the data collection phase of surveys.

Q.17 An increased interviewer training may minimise "________" error in a survey.

Q.18 The repeated call backs may ________ non-response error in survey.

Q.19 Sub-sampling of non-respondents, may minimise "________" error in a survey.

Q.20 Simplification of questionnaire, may minimise "________" error in a survey.

Q.21 A randomized response technique may "________" non-response error in survey.

Q.22 An incentive for co-operation can be utilized to "________" the response rate.

Q.23 In case of adjustment for "________" generally two types of substitution procedures are applied

Q.24 One of the "________" is the selection of a random substitute.

Q.25 Selection of a specially designated substitute is one of the "________"

Q.26 In a random substitution procedure, an "________" is selected to replace each non-responding unit.

Q.27 In specially designated substitute units, one or more "________" are identified to replace if necessary.

Q.28 A common procedure in dealing with non-response in a survey is to "________" a preassigned number of times.

Q.29 The call back estimates $\overline{y_i}$ (i = 1, 2, 3.....k) are "________" if the populations in different response classes have "________"

Q.30 Imputation is the process of explicitly or implicity substituting plausible data for "________" or "________" items in survey records.

Q.31 Imputation is the process of "________" or "________" substituting plausible data for incomplete or inconsistent items in survey records.

Q.32 In socio economic surveys, often an inefficient result may occure due to "________" arising out of "________" and non-response error.

Q.33 To tackle the problems of non-response in sample surveys, Warner (1956) introduced a "________".

20.3 National Sample Surveys

Q.34 National sample survey was initiated in 1950 to conduct sample surveys with a view to provide the "________" with the ________ data.

Q.35 The data collected by NSS can be utilized for planning for "________".

Q.36 NSS is a "________" which is carried out in the form of round.

Q.37 Each round of NSS covers the same topics of "________"

Q.38 Duration of each round of NSS varies from "________ to ________ months."

Q.39 In the first few rounds of NSS, emphasis was given to collect information for the computation of "________".

Q.40 To work out the national income, NSS collected "________" for the expenditure and "________" house hold interprises.

Q.41 In 8th round, the NSS, shifted to study the ________ of land holdings.

Q.42 In 9th round NSS, the primary object was to study the ________ & ________ situations in the country.

Q.43 In 10th round NSS, a study was made related to the estimation of ______ & ______ of cerial crops alongwith socio-economic enterprises.

Q.44 Statistical information about a place, region, state or nation is necessary for running ________ efficiently.

Q.45 The land revenue rate was based on ________ production and ________ of soil.

Q.46 The first census of India covering the whole of country was conducted in the year ________.

Q.47 In order to collect and compile economic data at state level, a ________ was established at the centre in 1895.

Q.48 An enquirey committee named as "________" was established in 1925 under the chairmanship of Sir. M.________

Q.49 Immediately after the report of "Economic Enquirey Committee" the report of "________" on agriculture was made out in 1928.

Q.50 "Statistical Research Bureau" was established in the year ________".

Q.51 An act known as "________". "Industrial statistics Act" was passed in the year ________.

Q.52 One wing of department of statistics of cabinet secretariate is the "________" set up in 1950.

Q.53 Central government established C.S.O. under ________.

Q.54 The objective of C.S.O. is to create co-ordination of large variety of ""________" collected at centre and ________ level.

Q.55 C.S.O. provides the advisory services related to statistical matter particulary to standardize the ________ and ________ to maintain uniformity throughout the country.

Q.56 C.S.O. collects the statistical data related to ________.

Q.57 C.S.O. compiles the ________.

Q.58 C.S.O. provides statistical data of nation to "________".

Q.59 C.S.O. attends the work of "________" held in India and abroad.

Q.60 C.S.O. provides statistical trainning to "________".

Q.61 C.S.O. co-ordinates the statistics collected by different ________ and ________.

Q.62 C.S.O. supplies information regarding methologies adopted in________.

Q.63 The "Directorate of Industrial Statistics" was set up in the year ________ at Kolkatta.

Q.64 C.S.O. lays down the definition and concepts in conformity with "________".

Q.65 Planning cell was set up in C.S.O. in the year ________ to maintain liaison between I.S.I and ________.

Q.66 All statistical matters related to planning are dealt within "________" of C.S.O.

Q.67 A National Income Unit (NIU) was created in the year ________ .

Q.68 The information of administrative and public interest is displayed through chart, ________ and ________ .

Q.69 N.S.S.O. stands for ________.

Q.70 NSSO was established in Jan., 1950 in the "Department of ________" ministry of ________ .

Q.71 NSSO provides methodology and analyses information which are useful to ________.

Q.72 NSSO provides information about new developments in the ________ and ________ activities.

Q.73 NSSO has the unified control of the governing council for survey ______ field-operation, data ________, economic ________ and publication of NSS data.

Q.74 At present, NSSO consists of ________ divisions with ________ at the apex.

Q.75 Survey design and research work is a division of ________.

Q.76 The executive officer of NSSO is the ________ of the governming council.

Q.77 The field operation division of NSSO with the 41 regions have been grouped into ________ zones.

Q.78 Each zone of NSSO has been placed under the charge of ________

Q.79 Enumeration of individuals under state direction at regular intervals is called "________".

Q.80 The latest population census has been carried in the year ________ .

Q.81 Defacto method or Date System, is the ________ .

Q.82 Dejure method or period system is the method to conduct ________.

Q.83 The first census of free India was conducted in the year ________.

Q.84 The census act of India was passed in the year ________.

Q.85 The census commissioner did the duties of "________" during the inter-census period.

Q.86 The distinction between "House" and "Household" was made first time in the year ________.

Q.87 House was defined as a dwelling place with a separate ________.

Q.88 Houlsehold was defined on the basis of "________".

Q.89 A group of persons living together and dinning from a common Kitchen were regarded as the member of same "________".

Q.90 The questionnare for 1961 census was prepared in the consultation of "___________".

Q.91 The questionnare of 1961 census consisted of two parts namely "________" and "________".

Q.92 The economic census survey was the responsibility of "_________".

Q.93 The 1981 economic census survey is the base of all the future "________"' and "________" programmes.

20.4 Agricultural Statistics System in India

Q.94 In ancient days, the records of agricultural and their categories was maintained on the basis of "________" and "________".

Q.95 Ancient rulars collected money in the form of "________".

Q.96 All the statistics having its impact on agricultural economy of the country, may be regarded as "________".

Q.97 The statistics related to land utilization, crop production, livestock and agricultural prices, poulty and forestry etc. comes under the category of "________".

Q.98 The agricultural statistics is mostly collected by the "________" and "________".

Q.99 Agricultural statistics is collected at centre as well as ________ level.

Q.100 Each state has a Directorate of "________ & ________".

Q.101 The total area statistics are maintained by sources "________" and village records maintained by "________".

Q.102 The records of area under crops is maintained by two sources (i) offical records based on "________" and (ii) NSS series based on "________".

Q.103 The land utilization statistics are published in "________".

Q.104 The individual states publish agricultural statistics in "________. and "________".

Q.105 Agricultural production includes the production of "________" and "______" crops.

Q.106 The output of agricultural production is estimated by multiplying the area under a crop by the "________"/hectare) in the season.

Q.107 Two methods to collect yield statistics of various crops are (i) "________"and (ii)Random sampling method.

Q.108 The traditional method of yield estimation, is known as "________".

Q.109 In case of yield estimation process, the crop condition is known as "________" or the "________".

Q.110 I.C.A.R. introduced the "________" in 1942 for cotton crop.

Q.111 For the crop yield estimation, the "________" is adopted for the "________" experiments.

Q.112 Now the crop cutting experiments are conducted by the "Directorate of ________ and ________".

Q.113 The NSSO conducts "________".on major cerial crops during course of their regular survey rounds.

Q.114 A detailed agricultural census in India was conducted in the year "________".

Q.115 The second agriculture census was conducted during the agriculture year "________".

Q.116 During the second agricultural census in India, one "________" and one "________" season was covered completely.

Q.117 Agricultural census (1980-81) is 3rd of its kind in India sponsered by "United Nations _____________".

Q.118 In India, agricultural census is conducted under the "Ministry of Agriculture" in "________" and the "Department of Revenue" in the "_________".

Q.119 Another agricultural survey named as "________" was conducted in the agricultural year (1981 -82).

Q.120 During the survey year (1981-82), information was collected in a "________" for both the seasons.

Q.121 The first cattle census on all India basis was conducted from December, ________ to April, ________.

Q.122 The livestock census is held quinquennially by the "Directorate of ________ and ________.

Q.123 Statistics regarding livestock and poultry, are published in "________" quinquennially.

Q.124 The information related to the "Volume of timber" and "round wood" are included in"________".

Q.125 The data related to area, the volume of standing timber are annually released in the "________" by D.E.S.

Q.126 For statewise major products, the data on quantities and royalty value are published annually in the "________".

Q.127 The fishery statistics is available only in published reports of the "________" from Madras & Bengal.

Q. 128 An estimate of output of fish are worked out separately for "________" and "________" at state level.

Q.129 The data on landing of marine fish are collected by the "________" Cochin.

Q.130 The National Income Committee, was set up by theGovt. of India in the year "________".

Q.131 The first results of "National Income" estimates were presented by the C.S.O. in "________".

Q.132 The "sum of employees compensation" and the net income from property and entrepreneurship" represents the "________" of the country.

Q.133 In simple words "The value of commodities and services produced during a particular period counted without duplication" is the"________.".

Q.134 National income estimate gives an idea of purchasing power of people in the "________".

Q.135 National income statistics provides a useful guide-line in the formulation of "________".

Q.136 National income statistics (NIS) is the basis to study the problems faced by economically in "________".

Q.137 NIS provides a basis for the future "________" of a country

Q.138 The three methods used to estimate the "________" are (i) output or production method, (ii) income method and (iii) expenditure method.

Q.139 The output method is also known as "________".

Q.110 Inventory method consistes in finding out the "________" of all goods and services provided by individuals and "business________".

Q.141 In income method, the national income consists of "________" all incomes by the way of wages, interest, rents & profits.

Q.142 In expenditure, the national income is equal to the sum of "________ and ________.".

Q.143 The best source of understanding the status of the country is "________".

Q.144 RBI has full-fledged department of"________ and ________".

Q.145 The data and information on banking, exchange, currency and finance etc. are published by "________".

Q.146 Financial statistics used by Govt. of India, Foreign Govt. and I.M.F. is compilled by "________".

Q.147 Statistical tables related to bonus in India, are published by "________".

Q.148 Reports on currency and finance (annual) are published by "________".

Q.149 Finance accounts of central and state's govt. are published by "_____".

SECTION-C : MULTIPLE CHOICE QUESTIONS AND ANSWERS

20.1 Sampling on Successive Occassions

Q.1 In applying varying probability sampling, the information on total x is required to know the selection probabilities as :

(a) P_i (i = 1,2, 3,..., N) (b) P_i (i = 1,2, 3,..., n)

(c) πi (i = 1,2, 3,..., N) (d) π_i (i = 1, 2, 3, ..., n)

Q.2 The technique of "double sampling" was first time formulated :

(a) J. Neyman (1938) (b) P.C. Mahalanobis (1943)

(c) R.A. Fisher (1950) (d) Cox (1945)

Q.3 When the sample for main surveys is selected in three or more phases, the procedure is termed as :

(a) Three phase sampling (b) Double sampling

(c) Multiphase sampling (d) Multistage sampling

Q.4 When the same population is sampled repeatedly, the opportunity for a flexible sampling design is :

(a) Increased (b) Decreased

(c) Not changed (d) None

Q.5 Sampling for two occassions was first time considered by :

(a) Murthy (1964) (b) Cochram, W.G. (1937)

(c) Jessen (1942) (d) Mahalanobis, P.C. (1951)

Q.6 In case of two occassions of surveys, one has a large flexibility in using :

(a) Sampling scheme (b) Estimate of attribute

(c) Sampling scheme & estimates (d) None

Q.7 When probability of happening of a rare attribute is very small, then sampling procedure applicable is :

(a) Inverse sampling (b) Double Sampling

(c) Sequential sampling (d) Both (a) & (c)

Q.8 Inverse sampling procedure is also termed as ;

(a) Double sampling (b) Sequential sampling

(c) Subsampling (d) Two phase sampling

20.2 Non-Sampling Errors

Q.9 Non-response may occure due to :

(a) Omission on part of investigator (b) Lapses on part of investigator

(c) Refusal from respondents (d) All the above

Q.10 Non-response may occure due to :

(a) Non-availability of respondent during survey

(b) Non-cooperation of interviewer

(c) Refusal of respondent
(d) Both (a) & (c)

Q.11 An item non-response may occure due to :
(a) Refusal of respondent to a specific question
(b) Refusal of investigator to ask a specific question
(c) Non-refusal of respondent to a specific question
(d) None

Q.12 For arising an "Error in measurement" the measuring device may be :
(a) Biased (b) Inaccurate
(c) Inconsistent (d) All the three

Q.13 An increased interviewer training, may :
(a) Minimise non-response error in survey
(b) Maximise non-response error in survey
(c) Increase the quality of data collected
(d) Both (a) & (c)

Q.14 Repeated call back may :
(a) Minimise non-response error
(b) Maximise non-response error
(c) Maximise the coverage of data collected
(d) Both (a) & (c)

Q.15 Sub-sampling of non-respondents, may :
(a) Minimise the non-response error
(b) Maximise the non-response error
(c) Maximise the coverage of survey work
(d) Both (a) and (c)

Q.16 Simplification of questionnaire contact, may:
(a) Maximise non-response error
(b) Minimise non-response error
(c) Maximise the coverage of surveyed data
(d) Both (b) & (c)

Q.17 An incentive for co-operation can be used to :

(a) Maximise response rate

(b) Minimise non-response error

(c) Maximise the coverage of data collected

(d) All the three

Q.18 In case of unit non-response, the substitution procedure applicable is:

(a) Selection of random substitute

(b) Selection of systeematic substitute

(c) Selection of specially designated substitute

(d) Both (a) & (c)

Q.19 In a random selection procedure, to replace each non-responding unit, an

(a) Additional unit is selected

(b) Extra unit is selected

(c) Some units from the sample, are repeated

(d) Both (a) & (b)

Q.20 The call back estimators $\overline{y}_i$ (i = 1, 2, ... , k)are biased if the populations in different response classes have :

(a) Different means (b) Same means

(c) Unbiased means (d) None

Q.21 In traditional sampling procedures, the socio-economic surveys often gives :

(a) An inefficient result (b) An efficient result

(c) A consistent result (d) None

Q.22 To tackle the non response problem in sample surveys, the randomised, response technique was introduced by :

(a) J.Neyman (b) Prof. R.A. Fisher

(c) Warner (d) None.

20.3 National Sample Surveys

Q.23 National sample surveys was initiated in the year :

(a) 1955 (b) 1951

(c) 1950 (d) 1948

Q.24 N.S.S. conducts sampling enquiries to provide to the government and other organisation, the :

(a) Socio-economic data (b) Demographic data

(c) Data related to health (d) Educational statisticis

Q.25 N.S.S. provides data on socio-economic problems to :

(a) Government (b) Non-government organisation

(c) Both (a) & (b) (d) None

Q.26 Data collected by NSS can be utilised for:

(a) Planning of national development

(b) Income tax assessment

(c) Planning of educational system in India

(d) Tackling health problems of country

Q.27 Data collected by NSS can be utilised for :

(a) National development (b) Various research purposes

(c) Assessment of income tax (d) Both (a) & (b)

Q.28 Each round of N.S.S. covers some topics of:

(a) Future interests (b) Past interest

(c) Current interest (d) Both (b) & (c)

Q.29 Duration of each round of N.S.S. varies from :

(a) (3-6) months (b) (3-8) months

(c) (3-7) months (d) (4-9) months

Q.30 First few rounds of NSS gave emphasis to collect information for the computation of:

(a) Income tax (b) National income

(c) Educational status of India (d) National health statistic

Q.31 To work out the national income, the NSS collects statistics for :

(a) Expenditure (b) Saving

(c) Small scale house hold enterprise (d) Both (a) & (c)

Q.32 In 8th round, NSS tried to study the :

(a) Distrubution of land holdings

(b) National policy for higher education

(c) National policy on primary health centres

(d) National policy to have land records

Q.33 During 9th round survey, N.S.S. caried out information to study the :

(a) Employment situation in country

(b) Unemployment situation in country

(c) Status of higher education in country

(d) Both (a) & (b)

Q.34 In 10th round of survey, the NSS made study related to :

(a) Acrage and yield rate of cerial crops

(b) Educational policy of the country

(c) Revenue from income tax

(d) Revenue from land utilisation by farmers

Q.35 In 10th round survey, NSS collected data to :

(a) Estimate the acrage and yield rate of cerial crop

(b) Socio-economic enterprise

(c) Estimate the birth and death rate in India

(d) Both (a) & (b)

Q.36 In (11 & 12)th round, NSS made a study on :

(a) Revenue received from land record in country

(b) Revenue received from income tax of service man

(c) Revenue received from corporate world

(d) Economic conditions of agriculture in India.

Q.37 The land revenue rate was based on...

(a) Agriculture production (b) Wheat production

(c) Soil fertility (d) Both (a) & (c)

Q.38 The first Census of India covering whole of the country was conducted in:

(a) 1885 (b) 1881

(c) 1981 (d) 1951

Q.39 In order to collect and compile economic data at state level, an organization established at centre in 1895 was :

(a) Statistical Buereau (b) Directorate of Economics

(c) NSSO (d) C.S.O.

Q.40 An enquiry committee named as ''Economic enquiry committee" was established in 1925 under the chairmanship of;

(a) P.C. Mahalanobis (b) Sir. M. Visvessraya

(c) P.V. Sukhatme (d) Prime Minister

Q.41 Immediately after the report of "Economic Enquiry Committee" the report of "Royal Commission" was made out in the year.

(a) 1925 (b) 1930

(c) 1928 (d) 1951

Q.42 Statistical Research Bureau, was established in the year :

(a) 1933 (b) 1930

(c) 1950 d) 1948

Q.43 An act known as "Industrial Statistics Act" was passed in the year :

(a) 1950 (b) 1942

(c) 1947 (d) 1945

Q.44 One wing of the "Deptt. of Statistics" of cabinet secretariate set up in 1950 is the:

(a) C.S.O. (b) N.S.S.O.

(c) Statistical Buereau (d) None

Q.45 Central Govt established C.S.O. under______

(a) Cabinate Secretariate

(b) Planning Commission

(c) Ministry of Statistics
(d) Ministry of Economics & Statistics

Q.46 The objective of C.S.O. is to create co-ordination of large variety of____
(a) Economics Information (b) Statistical Information
(c) Agricultural Information (d) Health Information

Q.47 C.S.O. provides services related to statistical matter particularly to standardize the:
(a) Concept (b) Definition
(c) Analysis (d) Both (a) & (c)

Q.48 C.S.O. collects statistical data on :
(a) Agriculture (b) Economics
(c) Planning (d) None

Q.49 C.S.O. Compiles the:
(a) National Income Estimate
(b) Health Data in the Country
(c) Agricultural data in the country
(d) Employment data of the country

Q.50 C.S.O. provides statistical data of Indian nation to :
(a) World Health Organization
(b) United Nations Statistical Offices
(c) Food & Agricultural Organization
(d) World Bank

Q.51 C.S.O. attends the works of "________" held in India & abroad :
(a) International confereence (b) National conference
(c) International work-shop (d) National work-shop

Q.52 C.S.O. coordinates the statistics collected by different "________".
(a) Ministries & N.S.S.O.
(b) Ministries & Planning commission
(c) Ministries & Directorate of Economics
(d) Ministries & Directorate of Statistics

Q.53 C.S.O. supplies information regarding methodologies adopted in...
(a) Analysis of data (b) Collection of data
(c) Compilation of data (d) Editing of data

Q.54 "Directorate of Industrial statistics" was set up in the year :
(a) 1950, at Calcutta (b) 1944, at Calcutta
(c) 1948, at Bombay (d) 1952, at Bombay

Q.55 C.S.O. lays down the definition and concepts in conformity with :
(a) National Standard
(b) International Standard
(c) Ministry of H.R.D.
(d) Ministry of Statistics & programme implementation

Q.56 I.S.I.C. stands for :
(a) Internantional Standard Industrial Classification
(b) Indian Standard Industrial Classification
(c) Indian Statistical Institute for Classification
(d) International Statistical Institute for Classification

Q.57 Planning cell in C.S.O. was set up in the year 1955 to maintain the liaison between I.S.I, and ......
(a) Planning commission (b) Ministry of Economics
(c) Ministry of Statistics (d) Ministry of H.R.D.

Q.58 All statistical matters related to planning are dealt within :
(a) Planning commission (b) C.S.O.
(c) Planning cell of C.S.O. (d) N.S.S.O.

Q.59 National Incocme Unit (NIU) was created in the year :
(a) 1951 (b) 1955
(c) 1950 (d) 1948

Q.60 NSSO stands for :
(a) National Sample Survey Organization
(b) National Sample Services Organization
(c) National Security of Survey Organization
(d) National Statistical Survey Organization

Q.61 NSSO was stablished in the department of:

(a) Economic Affair & Ministry of Finance

(b) Statistics & Ministry of Finance

(c) Economics & Finance

(d) Economics & Statistics

Q.62 NSSO provides information about new developments in :

(a) Economics & social affairs (b) Agriculture & Health affairs

(c) Health & Education services (d) Economic & Statistical field

Q.63 At present, NSSO consists of 4 divisions with :

(a) A chairperson at the apex

(b) Chief Executive officer at the apex

(c) Deputy Director at the apex

(d) Director General at the apex

Q.64 Sample survey design and research work is a division of;

(a) C.S.O.

(b) NSSO

(c) Planning commission

(d) Ministry of Statistics & Economics

Q.65 The executive officer of NSSO is the ______ of the governing council.

(a) Member Secretary (b) Deputy Director

(c) Joint Director (d) Director General

Q.66 The field operation division of N.S.S.O. with its 41 regions have been divided in to ______

(a) 7 zones (b) 4 zones

(c) 5 zones (d) 3 zones

Q.67 Each zone of NSSO has been placed under the charge of:

(a) Deputy Director (b) Joint Director

(c) Officer Incharge (d) Chief Statistical Officer

Q.68 Enumeration of individuals under state direction at regular interval is known :

(a) Sample survey (b) Population census

(c) Complete enumeration (d) Both (b) & (c)

Q.69 Defacto Method or Date System, is the methods of _______

(a) Population census (b) Complete enumeration

(c) Sample survey (d) Both (a) & (b)

Q.70 De-jure Method or period system is the method to conduct ________

(a) Census (b) Sample survey

(c) Complete enumeration (d) Both (a) & (c)

Q.71 The first census of free India was conducted in the year:

(a) 1948 (b) 1951

(c) 1950 (d) 1952

Q.72 The Census act of India was passed in the year :

(a) 1951 (b) 1950

(c) 1948 (d) 1952

Q.73 The Census Commissioner did the duties during the inter-censal period as_________

(a) Director General (b) Registrar General

(c) Chief of the census office (d) Officer-incharge

Q.74 The-distinction between "House & House-Hold" was made first time in the year :

(a) 1948 (b) 1950

(c) 1952 (d) 1951

Q.75 House-hold was defined on the basis of:

(a) Chulha

(b) Kitchen

(c) House with one main enterance

(d) Number of Kitchen in the house

Q.76 A group of persons living together and dinning from a common kitchen, were considered as the member of the same :

(a) House (b) House-hold

(c) Kitchen (d) None

Q.77 The questionnaire for 1961 census was prepared in the consultation of ________

(a) Planning Commission (b) N.S.S.O.
(c) C.S.O. (d) Ministry of Statistics

Q.78 The questionnare of 1961 census consisted of two parts named as :
(a) Household schedule & Individual slip
(b) Head of the household schedule & Individual slip
(c) Individual slip & house hold schedule
(d) Both (a) & (c)

Q.79 The economic census survey was the responsibility of:
(a) Registrar General of India
(b) Census Commisionor
(c) Chief of the Census Commission
(d) Director General of Census Commission

Q.80 The 1981 economic census survey is the base of all future ________
(a) Planning (b) Developmeent
(c) Education policy (d) Both (a) & (b)

20.4 Agricultural Statistics System in India

Q.81 In ancient days, the records of agricultural land and their categories were maintained on the basis of:
(a) Production (b) Soil fertility
(c) Irrigation facility (d) Both (a) & (b)

Q.82 Ancient rulars collected the money in the form of:
(a) Land Revenue (b) Production cost
(c) Area sown (d) Irrigated land

Q.83 All statistics having its impact on a gricultural economy of the country may be regarded as :
(a) Agricultural Statistics (b) Production Statistics
(c) Agricultural Economics (d) National Economics

Q.84 Agricultural Statistics is mostly collected by:
(a) Directorate of Statistics & Economics
(b) Directorate of Economics & statistics

(c) National Sample Survey Organization
(d) Ministry of Economics & Statistics

Q.85 Agricultural Statistics is collected at:
(a) Central level (b) State level
(c) Both (a) & (b) (d) District level

Q.86 Each state has a "Directorate" of:
(a) Economics (b) Statistics
(c) Agricultural production (d) Both (a) & (b)

Q.87 The records of area under crops in India is maintained by :
(a) Offical records based on village level
(b) NSS records based on sample survey
(c) Offical records based district level
(d) Both (a) & (c)

Q.88 The total area statistics in India is maintained by :
(a) Surveyor General of India (b) Revenue Deptt at village level
(c) Both (a) & (b) (d) NSSO of India

Q.89. Eand utilization statistics are published by :
(a) State Agricultural Statistics
(b) Indian Agricultural Statistics
(c) Revenue Deptt of India
(d) Indian Council ofAgril. Research

Q.90 Individual state publishes agricultural statistics in :
(a) Season Report (b) Crop Report
(c) Both (a) & (b) (d) Bulletin of Agril. Stats

Q.91 Agricultural production includes the production of:
(a) Food crops (b) Non food crops
(c) Both (a) & (b) (d) Rabi crops

Q.92 The method to collect yield statistics of various crops is :
(a) Traditional Method (b) Random Sampling Method
(c) Both (a) & (b) (d) None of the three

Q.93 The traditional method of yield estimation is known as :

(a) Annawary System (b) Crude method

(c) Oldest method (d) None

Q.94 In case of yield estimation process, the crop condition is known as :

a) Condition factor (b) Annawary system

(c) Both (a) & (b) (d) Yield factor

Q.95 I.C.A.R. introduced the estimation method in 1942 for cotton crop known as :

(a) Purposive method (b) Random sampling method

(c) Cluster sampling method (d) None

Q.96 For the crop yield estimation, the survey sampling design is the :

(a) Stratified random sampling design

(b) Simple random sampling design

(c) Multistage random sampling design

(d) Both (a) & (c)

Q.97 The crop cutting experiments are conducted by the

(a) Directorate of Economics (b) Directorate of statistics

(c) Both (a) & (b) (d) I.C.A.R.

Q.98 NSSO conducts ___________ on major cerial crop during their regular survey round :

(a) Crop cutting expts (b) Sample survey

(c) Purposive sample survey (d) Quota sampling survey

Q.99 A detailed agricultural census was conducted in the year:

(a) 1951-52 (b) 1970-71

(c) 1971-72 (d) 1974-75

Q.100 The second agricultural census was conducted during the agriculture year :

(a) 1973-74 (b) 1976-77

(c) 1979-80 (d) 1974-75

Q.101 During the second agriculture census, the time period covered completely was:

(a) One Rabi crop (b) One Kharif crop

(c) One Rabi & Kharif crop (d) Six months

Q.102 The agriculture census (1980-81) is 3rd of its kind in India sponsored by:

(a) Directorate of Economics & Statistics

(b) Ministry of Agriculture Govt. of India

(c) United Nations Food & Agriculture

(d) Ministry of Statistics & Programme Implementation

Q.103 In India, the agriculture census is conducted under :

(a) Ministry of Agriculture in Centre

(b) Department of Revenue in State

(c) Both (a) & (b)

(d) National sample survey organization

Q.104 Another agriculture survey conducted in agriculture year (1981-82) was named as :

(a) Input survey (b) Sample survey

(c) Crop survey (d) Yield survey

Q.105 During agriculture survey year (1981-82) the information was collected for both seasons in a :

(a) Single visit (b) Two visits

(c) Three visits (d) None

Q.106 The first cattle census on all India basis was conducted from :

(a) December, 1919 to April, 1920

(b) December, 1920 to April, 1921

(c) December, 1921 to April, 1922

(d) December 1922 to April, 1923

Q.107 The livestock census is held quinquennially, by the :

(a) Directorate of Agril. Engg. & Food Science

(b) Directorate of Agriculture & Statistics

(c) Directorate of Economics & Statistics

(d) Ministry of Agriculture at centre & state level

Q.108 Statistics regarding livestock and poultry, are published quinquennially in:

(a) Indian Forest Statistics

(b) Indian Livestock Census

(c) Indian Livestock Poultry Census

(d) State's Bulletine of Agril. Statistics

Q.109 Information regarding "volume of Timber" and "round wood" are included in:

(a) Agricultural Statistics

(b) Forest Statistics

(c) Report of Council of Agril. Research

(d) None

Q.110 Data on area, the volume of standing timber, are annually released in the :

(a) Indian Forest Statistics

(b) States Bulletin of Agricultural Statistics

(c) Forestry in India

(d) Annual Forest Report of India

Q.111 Fishery statistics is available only in published reports of:

(a) Deptt. of Fishery in Chennai & Bengal

(b) Deptt. of Agriculture in Chennai & Bengal

(c) Indian Council of Agriculture Research

(d) Indian Agricultural Research Institute

Q.112 An estimate of output of fish is worked out at state level for :

(a) Marine fish (b) Inland fish

(c) Both (a) & (b) jointly (d) Both (a) & (b) but separately

Q.113 The data on landing of marine fish are collected by :

(a) Central Marine Fishery Research Instt., Cochin

(b) Fishery Research Institute at Bengal

(c) Both (a) & (b)

(d) None

Q.114 National Income Committee, was set up by the Government of India in the year:

(a) 1948 (b) 1949

(c) 1952 (d) 1951

Q.115 The first results of''National Income' estimates was presented by C.S.O in:

(a) National Income Statistics (b) National Income Report

(c) Bulletin of National Income (d) Annual Report of C.S.O.

Q.116 The "Sum of employees compensation and the net income from property and entrepreneurship" represents the :

(a) National income of the country

(b) Total income of the corresponding state

(c) Total revenue collected for the nation

(d) Total budget of the country

Q.117 The value of commodities and services produced during the particular period counted without duplication, is the :

(a) National budget (b) National Income

(c) Total revenue of the country (d) None

Q.118 National income estimate gives an idea of purchasing power of people in:

(a) Country (b) State

(c) District (d) Family

Q.119 National income statistics provides a useful guideline to formulate the :

(a) Budget of the country (b) Budget of states

(c) Expenditure of the country (d) Needs of the country

Q.120 The basis to study the problems faced by economically underdeveloped country is the :

(a) National budget

(b) National income statistics

(c) Total revenue received by nation

(d) Total I.T. Collected

Q.121 For the future planning of the country, the base is provided by :

(a) National income statistics

(b) Total income tax collected

(c) Total revenue received by Govt. of India

(d) Total product of food & non-food crops

Q.122 The methods to estmate the national income are:

(a) Output or production method (b) Income method

(c) Expenditure method (d) All the three methods

Q.123 The output method to estimate the National Income is also known as :

(a) Inventory method (b) Explantory method

(c) Productive method (d) Revenue method

Q.124 Inventory method is consists of finding out the :

(a) Market values of all goods

(b) Services provided by individuals

(c) Business interprise

(d) All the three

Q.125 In income method, national income consists of adding together all incomes received from the sources :

(a) Way of wages (b) Interest

(c) Rent & profit (d) All (a), (b) & (c)

Q.126 In expenditure method, the national income is equal to the :

(a) Sum of consumptions & savings

(b) Sum of input & output of goods

(c) Sum of deposit & withdrawl in banks

(d) Deposite and its compound interest

Q.127 The best source of understanding the economic status of country is the:

(a) Fincial statistics

(b) Production of food and non-food crops

(c) Total revenue collected

(d) Total of all kinds of taxes collected

Q.128 RBI has full fledged department of:

(a) Research and statistics (b) Income tax and expenditure

(c) Input & output of items (d) Both (a) & (c)

Q.129 Data and information on banking exchange, currency and finance etc. are published by:

(a) CSO (b) NSSO

(c) R.B.I. (d) Both (a) & (c)

Q.130 Financial statistics compiled by RBI is utilized by :

(a) Govt of India (b) Foreign Govt.

(c) I.M.F. (d) All the three

Q.131 Statistical tables related to banks in India, are published by :

(a) SB1 (b) RBI

(c) C.S.O. (d) N.S.S.O.

Q.132 Reports on currency and finance (annual) are published by :

(a) CSO (b) N.S.S.O.

(c) RBI (d) SBI

Q.133 Finance accounts of Central and State Govt. are published by :

(a) SBI (b) RBI

(c) C.S.O. (d) N.S.S.O.

20.5 Key Answers

20.5.1 True/False

1	True	2	True	3	True
4	True	5	True	6	True
7	True	8	True	9	True
10	True	11	True	12	True
13	True	14	True	35	True
16	True	17	True	18	True
19	True	20	True	21	True
22	True	23	True	24	True

25	True	26	True	27	True
28	True	29	True	30	True
31	True	32	True	33	True
34	True	35	True	36	True
37	True	38	True	39	True
40	True	41	True	42	True
43	True	44	True	45	True
46	True	47	True	48	True
49	True	50	True	51	True
52	True	53	True	54	True
55	True	56	True	57	True
58	True	59	True	60	True
61	True	62	True	63	True
64	True	65	True	66	True
67	True	68	True	69	True
70	True	71	True	72	True
73	True	74	True	75	True
76	True	77	True	78	True
79	True	80	True	81	True
82	True	83	True	84	True
85	True	86	True	87	True
88	True	89	True	90	True
91	True	92	True	93	True
94	True	95	True	96	True
97	True	98	True	99	True
100	True	101	True	102	True
103	True	104	True	105	True
106	True	107	True	108	True
109	True	110	True	111	True
112	True	113	True	114	True
115	True	116	True	117	True
118	True	119	True	120	True
121	True	122	True	123	True
124	True	125	True	126	True
127	True	128	True	129	True
130	True	131	True	132	True

133	True	134	True	135	True
136	True	137	True	138	True
139	True	140	True	141	True
142	True	143	True	144	True
145	True	146	True	147	True
148	True	149	True	150	True
151	True	152	True	153	True
154	True	155	True	156	True
157	True	158	True	159	True
160	True	161	True	162	True
163	True	164	True	165	True
166	True	167	True	168	True
169	True	170	True		

20.5.2 Fill in the Blanks

1. J. Neyman
2. Multiphase Sampling
3. Sampling Design
4. Jessen
5. Sampling Scheme,Estimate
6. Samller
7. Small,/Inverse sampling
8. Not fixed
9. Sequential sampling scheme
10. Individual
11. Omission, Interviewer Respondent
12. Item non-responsendent
13. Respondent,Interviewer
14. False information
15. Error in Tnmeasurement
16. Non-response error
17. Non-response error
18. Minimise
19. Non-response error
20. Non-response error
21. Minimise
22. Maximise

23. Unit non-response
24. Substitution
25. Substitution procedures
26. Additional units
27. Back up units
28. Call back the unit
29. Biased, Different means
30. Incomplete, inconsistent
31. Explicitly, emplicitly
32. Missing data, undercoverage
33. Randomize Response Technique
34. Govt. Socio-economic data
35. National development
36. Continusing survey
37. Current interest
38. (4 to 8)
39. National Income
40. Expenditure, small scale
41. Land holding
42. Empoyment, Unemployment
43. Average,yield rate
44. Economic condition
45. Agriculture, Fertility
46. 1881
47. Statistical Bureau
48. Economic enquiry, Committee Visversaroya
49. Royal commission
50. 1933
51. 1942
52. NSSO
53. Cabinate Secreteriate
54. Statistical Information, State
55. Concepts, Definition
56. Planning
57. National Income Estimate
58. United Nations Statistical offices

59. International statistical institute
60. Statistical personnel
61. Ministry, NSSO
62. Collection of data
63. 1944
64. International standard
65. International Industrial classification
66. 1955, Planning commission
67. Planning cell
68. 1955
69. Graph; Diagram
70. National sample survey office
71. Economic affair, finance
72. Research worker
73. Economic, Social
74. Design, Processing, Analysis
75. Four Functional, Chief Executive officer
76. NSSO
77. Member Secretary
78. Five
79. Deputy Director
80. Population census
81. 2010
82. Method of census
83. Census
84. 1951
85. 1948
86. Registrar General
87. 1951
88. Main enterance
89. Chulha
90. House-hold
91. Planning commission
92. House-hold schedule, Individual slip
93. Registrar General of India
94. Planning, Development

95. Fertility, Prodouction
96. Land Revenue
97. Agricultural Statistics
98. Agricultural Statistics
99. Directorate of economics, Statistics
100. State
101. Economics, Statistics
102. Surveyor General of India, Revenue Deptt.
103. Village Records, Sample Survey
104. Indian Agricultural Statistics
105 Season, Crop Report
106. Food, non-food
107. Average expected yield
108. Traditional Method
109. Annawari System
110. Conditional factor or Annawari statistics
111. Random Sampling Method
112. Stratified multistage random sampling design, Crop cutting
113. Economics, Statistics
114. Crop cutting experiment
115. (1970-71)
116. (1976-77)
117. Kharif, Rabi
118. Food and Agril. Organization
119. Centre, State
120. Input survey
121. Single visit
122. (1919-20)
123. Economics, Statistics
124. Indian Livestock Census
125. Forest statistics
126. Indian Forest Statistics
127. Forestry in India
128. Department of Fishary
129. Marine Inland
130. Centre Marine Fishary Research Institute

131. 1949
132. National Income Statistic
133. National Income
134. National Income
135. Country
136. Budget of the country
137. Underdeveloped, country
138. Planning of country
139. National Income
140. Inventory Method
141. Market value Enterprise
142. Adding together
143. Consumption, Savings
144. Financial Statistics
145. Research, Statisitics
146. RBI
147. RBI
148. RBI
149. RBI
150. RBI

20.5.3 Multiple Choice Questions and Answers

1.	a)	2.	a)	3.	c)
4.	a)	5.	c)	6.	c)
7.	d)	8.	b)	9.	d)
10.	d)	11.	a)	12.	d)
13.	d)	14.	d)	15.	d)
16.	d)	17.	d)	18.	d)
19.	d)	20.	a)	21.	a)
22.	a)	23.	c)	24.	a)
25.	c)	26.	a)	27.	d)
28.	c)	29.	b)	30.	b)
31.	d)	32.	a)	33.	d)
34.	a)	35.	d)	36.	d)
37.	d)	38.	b)	39.	a)
40.	b)	41.	c)	42.	a)

43. b)	44. b)	45. a)
46. b)	47. d)	48. c)
49. a)	50. b)	51. a)
52. a)	53. b)	54. b)
55. b)	56. a)	57. a)
58. c)	59. b)	60. a)
61. a)	62. b)	63. b)
64. b)	65. a)	66. c)
67. a)	68. d)	69. d)
70. d)	71. b)	72. c)
73. b)	74. d)	75. a)
76. b)	77. a)	78. d)
79. a)	80. d)	81. d)
82. a)	83. a)	84. b)
85. c)	86. d)	87. d)
88. c)	89. b)	90. c)
91. c)	92. c)	93. a)
94. c)	95. d)	96. a)
97. c)	98. a)	99. b)
100. b)	101. c)	102. c)
103. c)	104. a)	105. a)
106. a)	107. c)	108. b)
109. b)	110. a)	111. a)
112. d)	113. b)	114. b)
115. a)	116. a)	117. b)
118. a)	119. a)	120. d)
121. a)	122. d)	123. a)
124. d)	125. d)	126. a)
127. a)	128. a)	129. c)
130. d)	131. b)	132. c)
133. b)	134. b)	

Bibliography

Cochan W.G. (1946). Relative Accuracy of Systematic and Stratified Random Sampling for a Certain Class of Populations. Ann. Math. Statist, 17, 164-177.

Cochan, W.G. (1977). Sampling Technique, 3rd edition, Wiley Eastern Ltd., New Delhi.

Cochran, W.G. and Cox , G.M. (1957). Experimental Designs., 2nd edition. John Wiley & Sons, New York.

Drapper, N.R. and Smith, H. (1966). Applied Regression Analysis, Wiley series in Probability & Mathematical statistics, 2nd edition.

Fisher, R.A. (1947). The Design of Experiments., Oliver and Boyed, Edinburgh, 4th edition.

Federer, W.T. (1967). Experimental Design, Theory and Application. Oxford and I.B.H. Publishing Co. Bombay.

Gupta, H.C. and Kapoor, V.K. (1990). Fundamental of Applied Statistics, 3rd edition, S. Chand and Sons, New Delhi.

Hanson, M.H., Harwitz, W.N. and Medow, W.G. (1953). Sample Survey Methods and Theory, John Wiley and Sons, New York.

Kushwaha, K.S. & Kumar, Rajesh (2009). The Theory of Sample Surveys and Statistical Decisions, 1st edition, New India Publishing Agency, New Delhi.

Kushwaha, K.S. & Kumar, Rajesh (2009). Basic Concepts in Statistics, 1st edition, New India Publishing Agency, New Delhi.

Mood, A.M., Craybill, F.A. & Bose, D.D. (1974). Introduction to the Theory of Statistics, McGraw Hill Kogakusha, Ltd, London.

Murthy, M.N. (1964). Product Method of Estimations, Sankhya A. 26, 69-74.

Murthy, M.N. (1977). Sampling Theory and Methods, Statistical Publishing Society, Calcutta.

Panse, V.G. & Sukhatme, P.V. (1978). Statistical Methods for Agricultural Workers., I.C.A.R., New Delhi.

Singh, D. and Chaudhary, F.S. (1989). Theory and Analysis of Sample Survey Designs, Wiley Eastern Ltd. New Delhi.

Steel, R.G.D., and Torre, J.H. (1980). Principles and Procedures of Statistics. A Biometrical Approach, 2nd ed., McGraw Hill Kogakusha, Ltd. (International Student Edition).

Sukhatme, P.V. and Sukhatme, B.V. (1970). Sampling Theory of Survey with Applications. Asia Publishing House, New Delhi.

Sukhatme, *et. al.* (1984). Sampling Theory of Surveys with Applications, IOWA, State University Press, Ames, Iowa (U.S.A.) and Ind. Soc. Agril. Stats, New Delhi.

United Nations (1949). Recommendations Concerning the Preperation of Reports of Sampling Surveys, Statistical Series C, No.1 New York, reprinted in Sankhya. 9, 392-398.

22

Model Papers

Model Test Paper 1

1. Which of the following states has nutrient use of < 100 kg/ ha?
   (a) Madhya Pradesh
   (b) Karnataka
   (c) West Bengal
   (d) Uttar Pradesh
2. Which of the following is the leading state in maize production
   (a) Bihar
   (b) Uttar Pradesh
   (c) Punjab
   (d) Gujarat
3. Sexual dimorphism means
   (a) Males and females are separate
   (b) Females possess didelphic ovaries
   (c) Males and females are morphologically dissimilar
   (d) None of the above
4. Which of these statements regarding RFLP analysis is correct?
   (a) RFLP analysis requires Southern blotting for detection of fragments
   (b) RFLPs can identify single base pair changes at any site in the chromosome
   (c) An RFLP typically produces several different alleles.
   (d) All of these are correct
5. Who defined "economics is the science which treats of wealth"?
   (a) F.A.Walker
   (b) J.B. Says
   (c) Adam Smith
   (d) D. Ricardo
6. "The additional benefit a person derives from a given increase of his stock of anything diminishes with the growth of the stock he has" is termed as ________.
   (a) LDMU
   (b) LEMU
   (c) Indifference curve
   (d) Revealed preference theory

7. The extension and contraction of demand occurs due to ________.
   (a) Change in price of the commodity
   (b) Shift in the demand curve
   (c) Change in taste and preferences of the consumer
   (d) Change in taste and preferences of the producer
8. Which word is originated from the Latin word Personare which used to mean the voice of an actor speaking through a mask?
   (a) Performance (b) Politics
   (c) Personality (d) Perception
9. Sequence Of Transformations Undergone By Sulfur Where It Is Taken Up By Living Organisms, Transformed Upon Death And Decomposition Of The Organism, And Converted Ultimately To Its Original State Of Oxidation
   (a) Sulfur Cycle (b) Carbon Cycle
   (c) Phosphorus Cycle (d) None
10. 'Degree to which the data predict the candidate's success as a manager'- is denoted by which of the following ?
   (a) Validity (b) Reliability
   (c) Honesty (d) Integrity
11. The United States is considered which type of society?
   (a) Haptic (b) Chronemics
   (c) Monochronic (d) Polychronic
12. Gyandoot project is an ICT initiative by which state?
   (a) Rajasthan **(b) Madhya Pradesh**
   (c) Gujarat (d) Haryana
13. Which one of the following was first published periodical on agriculture in India ?'
   (a) Agriculture today (b) Farming Today
   (c) Agriculture farming trends **(d) Bengal Agriculture Gazette**
14. In which of the following programme, 'Lady Village Extension Workers' (LVEW) under their circle imparted training in 43 skills of agriculture and allied fields ?
   (a) TANWA programme (b) WYTEP programme
   (c) TEWA programme (d) MAPWA programme

15. Female science graduates were recruited and given a one-year diploma training in agriculture at Jabalpur Agricultural University under which of the following programme ?
    (a) TANWA programme (b) WYTEP programme
    (c) TEWA programme **(d) MAPWA programme**
16. Explicit costs are also known as ____.
    (a) Implicit costs (b) Cash costs
    (c) Paid out costs **(d) Both b & c**
17. What type of relationship between exists between MC and MPP ?
    (a) Direct (b) Positive
    (c) Indirect (d) None of these
18. At the point of market equilibrium ___.
    (a) Supply < Demand **(b) Supply = Demand**
    (c) Supply e" Demand (d) Supply > Demand
19. Which one is not true ?
    (a) Enterprise is a product (b) Entrepreneurship is a process
    (c) Entrepreneur is a person **(d) None of these**
20. Which country has higher rank in GDP per capita ?
    (a) Luxerberg (b) Denmark
    (c) Canada (d) Switzerland
21. Dependence effect' was associated with _____.
    (a) Robertson **(b) Galbraith**
    (c) JM Keynes (d) AC Pigou
22. Macroeconomics is the branch of economics that deals with ____.
    (a) The economy as a whole
    (b) Imperfectly competitive markets
    (c) Only the long run adjustments to equilibrium in the economy
    (d) Individual industries and the behaviour of individual decision
23. All currencies other than the domestic currency of a given country are referred ____.
    (a) Reserve currencies (b) Near monies
    (c) Foreign exchange (d) Hard currency

24. In the terminology of economics and money demand, the terms M3 and M4 are also known as ___.

    (a) Short money (b) Long money

    (c) Broad money (d) Narrow

25. By increasing the 'bank rate', the RBI can:

    (a) Provide incentives to commercial banks to lend more to public

    (b) Provide incentives to commercial banks to lend less to public

    (c) Increase the money supply in the market

    (d) None of these

26. According to the effective demand principle

    (a) At a certain price, the output shall not be determined by any known factor

    (b) At a certain price, the output will remain unaffected by rise or fall in demand

    (c) At a certain price, equilibrium output will be solely determined by the aggregate demand

    (d) None of the above

27. The Phillips curve describes the relationship between ___.

    (a) The federal budget deficit and the trade deficit

    (b) Savings and investment

    (c) The unemployment rate and the inflation rate

    (d) Marginal tax rates and tax revenues

28. Non-homogenous production function is related to ____.

    (a) Return to scale

    (b) Constant return to scale

    (c) Law of return

    (d) Either increasing or decreasing return to scale

29. Fixed resources are abundant in which stage of production ?

    (a) First (b) Second

    (c) Third (d) None of these

30. Better use of land, better management and marketing are the features of___________.
    (a) Ranching (b) Diversified farming
    (c) Mixed farming (d) Specialized farming
31. On the upper ridge line MRTS equal to
    (a) Zero (b) Positive
    (c) Infinite (d) Negative
32. According to cost concept of sen committee, Cost A_2 equal to CostA_1 plus
    (a) Rent paid for leased-in land (b) Rental value of owned land
    (c) Interest on fixed capital (d) Imputed value of family labour
33. Which is not a use of production function ?
    (a) Biological production function on the basis of experimental data is useful for policy making
    (b) Farm production function is used for decision making on the farm level
    (c) Measure the relative contribution of inputs in total output singly as well as jointly
    (d) All of these
34. For maximum level of outputs is
    (a) MPP=0 (b) TPP=Max
    (c) MPP=Max **(d) Both a & b**
35. The commodity that was mostly stored in cold storage structures in India is :
    (a) Potato (b) Fish
    (c) Meat (d) Milk
36. In India, the value addition in food is nearly ____ as against 23 per cent of China.
    (a) 07% (b) 25%
    (c) 45% (d) 75%
37. Commodity futures markets came into existence in India since–
    (a) 1920 **(b) 1952**
    (c) 1921 (d) 1960

38. National Institute of Agricultural Marketing (NIAM) is located at –

    (a) Jaipur (b) Hyderabad

    (c) Bangalore (d) Cochin

39. The term issue price means at which :–

    (a) Buffer stock is given to public

    (b) FCI issues food grains to PDS

    (c) Government procures buffer stock

    (d) Buffer stock is maintained by FCI

40. Hewlett Packard has all in one machines for print, copy and scan, it is the example of which type of marketing –

    (a) Niche marketing (b) Target marketing

    (c) Grassroots marketing (d) Combo marketing

41. Primary Agricultural Cooperative Societies (PACS) working at __________

    (a) Village level (b) District level

    (c) State level **(d) All of these**

42. A firm is producing standardized products and enjoying the economies of scale by producing more. Which one strategy is following by firm ?

    (a) Differentiation strategy

    (b) Cost leadership strategy

    (c) Market dominance strategy

    (d) Market segmentation strategy

43. The process of sorting of produce into different lots having same characteristics with respect to quality specifications.

    (a) Assembling (b) Standardization

    (c) Grading (d) Storage and processing

44. Which type markets involve decisions such as which country to enter in, how to enter, how to adopt their product and services, and how to price ?

    (a) Consumer markets (b) Business markets

    (c) Global markets (d) Government markets

45. Trade allowances, sample products, varying terms of credit, and discounts are the part of which element marketing mix ?

    (a) Product (b) Price

    (c) Place **(d) Promotion**

46. The state having highest pesticide usage is

    (a) Maharashtra (b) Andhra Pradesh

    (c) Gujarat (d) Tamil Nadu

47. Creating charge against immovable property which includes land, buildings or anything that is attached to the earth, it is a case of _____.

    (a) Hypothecation (b) Pledge

    (c) Mortgage (d) None of these

48. International Monetary Fund (IMF) was established in the year ___________.

    (a) 1944 **(b) 1945**

    (c) 1948 (d) 1950

49. DICGC provides which of the following _____.

    (a) Insurance of bank deposits (b) Stability in banking system

    (c) Faith of public in banking system **(d) All of these**

50. The term "SWOT" used for to _____.

    (a) Analyse business opportunities

    (b) Think on market share strategy

    (c) Know competitor's market strategies

    (d) None of these

Model Test Paper 2

1. Which of the following is the discount rate that equates the present values of the two sets of flows ?

   (a) Discounted cash flow (b) Payback period

   (c) Discounted cash flow **(d) Internal rate of return**

2. Single window system in cooperative credit structure was introduced by:–

   (a) Mohan Kanda (b) Raiffiesen

   (c) Narasimhan (d) Nicholson

3. Which of the following are the benefits of co-operative organization ?

   (a) Economic benefits (b) Social benefits

   (c) Educational benefits **(d) All of these**

4. According to Harrod a technology change is neutral when _____.

   (a) Decreasing capital labour ratio

   (b) Increasing capital labour ratio

   (c) Constant capital labour ratio

   (d) All of these

5. Which growth theory was developed as reaction to omission and deficiency in the solow-swan new classical growth model ?

   (a) Dualistic theory **(b) The new endogenous**

   (c) Steady state growth (d) The big push theory

6. The null hypothesis is the _______.

   (a) Claimed value for the population parameter

   (b) Statement whose validity we would like to establish with reasonable certainty

   (c) True statement in a hypothesis test

   (d) False statement in a hypothesis test

7. The affected leaves are inverted boat; internodes become shorter in Chillies due to the infestation of

   (a) *Scirtothrips dorsalis* (b) *Trialeurodes ricini*

   (c) *Polyphagotarsonemus latus* (d) *Aleurolobus barodensis*

8. If you reject a joint null hypothesis using the F-test in a multiple hypothesis setting, then

   (a) A series of t-tests may or may not give you the same conclusion.

   (b) The F-statistic must be negative

   (c) All of the hypotheses are always simultaneously rejected

   (d) The regression is always significant

9. The RESET test is used mainly to check for _____.

   (a) Collinearity (b) Orthogonality

   (c) Functional form (d) Capitalization on chance

10. The rationale behind the Koyck distributed lag is that it _____.

    (a) Eliminates bias

    (b) Increases the fit of the equation

    (c) Exploits an information criterion

    (d) Incorporates more information into estimation

11. If regression analysis is used to estimate the linear relationship between the natural logarithm of the variable to be forecast and time, then the slope estimate is equal to

    (a) The linear trend

    (b) The natural logarithm of the rate of growth

    (c) The natural logarithm of one plus the rate of growth

    (d) The natural logarithm of the square root of the rate of growth

12. The following assumption of LP problem is also called as the deterministic assumption

    (a) Proportionality (b) Divisibility

    (c) Non negetivity **(d) Certainty**

13. The unit cost equal to the increase in profit to be realized by one additional unit of the resource is termed as _____.

    (a) Price **(b) Shadow price**

    (c) Marginal factor cost (d) None of these

14. When making a location decision at the country level, which of these would be considered ?
    (a) Corporate desires
    (b) Land/construction costs
    (c) Air, rail, highway, waterway systems
    (d) Location of markets

15. Which of the following is a location analysis technique typically employed by a service organization ?
    (a) Purchasing power analysis (b) Linear programming
    (c) Queuing theory (d) Crossover charts

16. The transportation technique or simplex method cannot be used to solve the assignment problem because of ______.
    (a) Degeneracy (b) Non-degeneracy
    (c) Square matrix (d) All of these

17. The objective of assignment problem is to assign ______.
    (a) Number of origins to equal number of destinations at minimum cost
    (b) Number of origins to equal number of destination at maximum cost
    (c) Only to maximize cost
    (d) Only to maximize the profit

18. This is a method of finding out the best course of action for a firm in view of the anticipated countermoves from the competing organizations.
    (a) Game theory (b) Decision theory
    (c) Queuing theory (d) Markov process

19. If the sum of the game is not zero but the sum of the payoffs to both players in each cas e is constan t, then the game is refe rred as _____.
    (a) Constant sum games (b) Zero-sum game
    (c) Two person game (d) n-person game

20. In two person zero sum game ______.
    (a) Gain by one player is equal to loss by the other player
    (b) Two players gains
    (c) Gain by one player is more than the loss by the other player
    (d) None of these

21. The expected waiting time in the system is 10 minutes, and expected waiting time in the queue is 5 minutes then the service rate is ________.

(a) 1 /10 **(b) 1/5**

(c) 5 (d) 0

22. What is the Environmental Economics ?

(a) The branch of economics that studies how environmental and natural resources are developed and managed

(b) The branch of economics that shows how to exploit natural resources as quickly as possible

(c) The psychological study of relationships between humans and natural resources

(d) All of these

23. If government finds an endangered species living on your property, then____________.

(a) Owner has severe restrictions on their property

(b) The government in effect takes the property, because it severely limited the owner's choices

(c) Government does not compensate to protect the endangered species

(d) All of these

24. What kind of market failure is it, if a company producing medicines also pollutes the air ?

(a) Asymmetric Information (b) Monopolies

(c) Negative Externality (d) Open access property problem

25. The EPA was going to shutdown the Robbins Company, because it discharged too much polluted water. However, this caused the company's engineers to develop new methods to purify water, allowing the company to produce higher quality for a lower cost. Which idea explains this behaviour ?

(a) Porter Hypothesis (b) Global warming potential

(c) Coase Theorem (d) A Pigouvian Price

26. _________ involves bargaining between those who perpetrate pollution and those who suffer from such pollution, with 'property rights' determining the outcome of that bargain

(a) First Law of Thermodynamics (b) Pigouvian

(c) Existence value **(d) Coase Theorem**

27. Which country is the second largest polluter in the world ?

(a) USA (b) Great Britain

(c) Canada (d) India

28. Which of the following is correct ?

(a) Strong sustainability assumes that little or no substitution between physical and natural capital is possible.

(b) Environmental sustainability means that the resource use by previous generations should not exceed a level that would prevent subsequent generations from achieving a level of wellbeing at least as great.

(c) Weak sustainability means the value of the remaining stock of natural capital should not decrease.

(d) Environmental sustainability means the physical flows of aggregate resources should be maintained.

29. Maximum sustainable yield (MSY) in a fishery is ______.

(a) The efficient rate of catch of fish that maximizes the rent of fish stock

(b) The efficient rate of catch of fish that maximizes the amount of sustainable catch

(c) The maximum rate of catch of fish that is sustainable, but not necessarily efficient

(d) The maximum rate of catch of fish that is sustainable and yields maximum profit.

30. Which is basic and fundamental function of management ?

(a) Organizing **(b) Directing**

(c) Planning (d) Staffing

31. Which Fact Is Not True About Marketing ?

(a) Social Process (b) Create Utility

(c) Human Process **(d) Ownership transfer**

32. Who do not show initiative in visualising and implementing new ideas and innovations wait for some development which would motivate them to initiate unless there is an imminent threat to their very existence __________.

(a) Small scale **(b) Medium Scale**

(c) Large scale (d) None of these

33. The development of the agro industry can help ______.
    - (a) Create employment
    - (b) Increase standard of living
    - (c) Stabilize the economy
    - **(d) Make less lucrative agriculture**

34. Cholesterolemia means
    - **(a) Lack of functional LDL receptors**
    - (b) Lack of functional HDL receptor
    - (c) High sensitivity to fatty food intake
    - (d) None of the above

35. Median bulb in plant parasitic nematodes is a
    - (a) Secretory
    - (b) Glandular
    - (c) Globular
    - **(d) Muscular pumping organ**

36. Biochemical oxygen demand means
    - **(a) Amount of oxygen taken up by the microorganisms present in the water**
    - (b) Amount of oxygen released by the microorganisms present in the water
    - (c) Amount of nutrients taken up by the microorganisms present in the water
    - (d) None of these

37. A tree which had been proven to be genetically superior by means of progeny testing is known as
    - (a) Plus tree
    - (b) Check tree
    - **(c) Elite tree**
    - (d) Superior tree

38. In packing and storage of dried fruits and vegetables preventive and curative treatments are followed to avoid infestation. The two prominent curative treatments are
    - **(a) Thermal and fumigation**
    - (b) Desiccation and fumigation
    - (c) Thermal and preventive packing
    - (d) All of these

39. If the null hypothesis Ho in ANOVA is declared significant then theoretically the value of F statistic becomes
    - (a) Equal to 1.0
    - **(b) Greator than 1.0**
    - (c) Less than 1.96
    - (d) Greator than 1.96

40. The number of basic principle followed in L.S.D. is equal to -

(a) 1 (b) 2

(c) 3 (d) 4

41. The designs R.C.B.D. and L.S.B., both are

(a) Balanced designs (b) Connected designs

(c) Orthogonal designs **(d) All the three**

42. The theory of ANCOVA makes use of concepts of both technique of

(a) ANOVA and Principle least square

(b) ANOVA and Regression analysis

(c) ANOVA and Sub plot

(d) Regression and Sub plot

43. As per FSS Act 2006, Misbranded means an article of food if

i) False, misbranding or deceptive claims either on the label of the package, or through advertisement,

ii) The article is sold as an imitation of, or is a substitute for, or is likely to deceive,

iii) The package bears statement, design or device regarding the ingredients or the substances contained therein, which is false or misleading

iv) Manufacturers name and address is false or

v) Contains any artificial flavouring, colouring or chemical preservative which of the following is correct pair to define misbranded

(a) (i, ii, iii, iv, v) (b) (i, ii, iii, v)

(c) (i, ii, iv, v) **(d) (i, ii, iii, iv)**

44. Stoke's law states that the velocity of sphere will rise or fall in a liquid varies as the

(a) Diameter **(b) Square of its diameter**

(c) Radius (d) Sqare if its radius

45. The most accurate means of measuring ingredient is _______.

(a) Volume **(b) Weight**

(c) Number (d) Pressure

46. Gestation period of elephant is:

    (a) 310 days (b) 250 days

    (c) 150 days **(d) 630 days**

47. The state producing maximum milk:

    (a) U.P. (b) Bihar

    (c) Rajsthan (d) Punjab

48. The _________ structure of molecules such as proteins and fatty acids remains intact on application of high hydrostatic pressure.

    (a) Primary (b) Secondary

    (c) Tertiary (d) None of these

49. In chronological classification data are classified on the basis of _____________

    (a) Time (b) Attributes

    (c) Religion (d) None of these

50. Tick the correct statement

    (a) Respiration is a catabolic process in which breakdown of carbohydrates only takes place

    (b) Respiration is a catabolic process in which breakdown of carbohydrates, fats, proteins takes place

    (c) Respiration is a catabolic process in which breakdown of carbohydrates,fats, proteins takes place in the presence of certain enzymes and energy is released

    (d) Respiration in plants takes place through leaves only

Model Test Paper 3

1. Choose the correct answer. A thermodynamic state function is a quantity
   (a) Used to determine heat changes
   (b) Whose value is independent of path
   (c) Used to determine pressure volume work
   (d) Whose value depends on temperature only
2. Glycogen synthase is characterized by all of the following statements except
   (a) The enzyme exists in active and inactive forms
   (b) Uridine diphosphate glucose is a substrate
   (c) It is activated by phosphorylation
   (d) It requires a primer strand of glycogen
3. Which of the following statements about a lethal allele is NOT correct?
   (a) Lethal alleles are always recessive
   (b) Lethal alleles may have a late age of onset
   (c) Lethal alleles may be caused by mutations in essential genes
   (d) Lethal alleles may affect one individual differently than another
4. Consider three recessive alleles (a, b, and c) in a hypothetical organism. Assume that a female heterozygous for all three traits is testcrossed. Which of the data sets below would indicate that all three genes are linked on the same chromosome?
   (a) a+ b+ c+ = 378; a b c = 375; a+ b c = 370; a b+ c+ = 372; a+ b+ c+ = 370; a b c = 379; a+ b c+ = 380; a b+ c = 376
   (b) a+ b+ c+ = 940; a b c = 932; a+ b c = 282; a b+ c+ = 286; a+ b+ c+ = 245; a b c = 240; a+ b c+ = 40; a b+ c = 35
   (c) a+ b+ c+ = 940; a b c = 932; a+ b c = 125; a b+ c+ = 130; a+ b+ c+ = 938; a b c = 943; a+ b c+ = 132; a b+ c = 135
   (d) a+ b+ c+ = 300; a b c = 900; a+ b c = 300; a b+ c+ = 300; a+ b+ c+ = 300; a b c = 300; a+ b c+ = 300; a b+ c = 2
5. All the following statements about carnitine are true except
   (a) It can be synthesised in the human body
   (b) It can be synthesized from methionine and lysine
   (c) It is required for transport of short chain fatty acids into mitochondria
   (d) Its deficiency can occur due to haemodialysis

6. Which of the following statements regarding the regulation of trp operon expression by attenuation is correct ?

   (a) Rapid translation of the leader peptide prevents completion of mRNA transcript

   (b) Rapid translation of the leader peptide allows completion of mRNA transcript

   (c) The leader peptide sequence encodes enzymes required for tryptophan synthesis

   (d) The leader peptide sequence contains no tryptophan residues

7. Protein A will fold into its native state only when protein B is also present in the solution. However protein B can fold itself into native confirmation without the presence of protein A. Which of the following is true ?

   (a) Protein B serves as precursor for protein A

   (b) Protein B serves as molecular chaperon for protein A

   (c) Protein B serves as ligand for protein A

   (d) Protein B serves as structural motif for protein A

8. Choose the incorrect statement about peptide bond ?

   (a) Stable **(b) Charged and polar**

   (c) Trans in configuration (d) Partial double bond charater

9. The correct statement concerning RNA and DNA polymerases is

   (a) RNA polymerase use nucleoside diphosphates

   (b) RNA polymerase require primers and add bases at 5' end of the growing polynucleotide chain

   (c) DNA polymerases can add nucleotides at both ends of the chain

   (d) All RNA and DNA polymerases can add nucleotides only at the 3'end of the growing polynucleotide chain

10. RNA polymerase recognizes sites on DNA known as promoters. All of the following statements regarding these sites are correct except

   (a) A sequence of DNA base-pairs known as the pribnow box is located 5-10 bases upstream from the first base to be copied into RNA

   (b) All pribnow box sequences are variants of the sequence TATAATG

   (c) A second sequence, the -35 sequence, in many promoters, is located about 35 bases downstream from the pribnow box

   (d) The σ subunit of prokaryotic RNA polymerase is involved in recognition of the promoter sequence

11. Which of the following is not a gustatory lingual papilla ?

(a) Filiform (b) Lenticular

(c) Fungiform (d) Foliate

12. Which one of the following statements correctly describes initiation of protein synthesis in *E. coli*?

(a) The initiator tRNA binds to the Shine-Dalgarno sequence

(b) Three initiation factors are involved and IF2 binds to GTP

(c) The intermediate containing IF1, IF2, IF3 initiator tRNA and mRNA is called the 30S initiation complex

(d) Binding of the 50S subunit releases IF1, IF2, GMP and PPi

13. Which of the following statements is false for the nitric oxide gas?

(a) An intracellular signaling molecule

(b) Deamination of histidine results into nitric oxide production

(c) Stimulates guanylyl cyclase to produce cGMP

(d) Can be produced by activated neutrophils

14. Sweating sickness is cattle is mostly due to infestation of-

(a) Hyaloma spp (b) Dermacenter spp

(c) Rhephicephalus spp (d) Hemophysalis spp

15. What is the tail fin, the most important of all the median fins, also known as?

(a) Adipose fin (b) Anal fin

(c) Caudal fin (d) None

16. Following statement is true regarding rectal administration of drugs: -

(a) It is a convenient mode of the administration in presence of vomiting

(b) Rectal administration of drugs may produce systemic effects

(c) This bypasses chemical reactions at upper part of intestine

(d) All of the above

17. True statement in relation to dextio isomer of 3 hydroxy N-methyl morphinan is: -

(a) It is an antitussive and lacks analgesic activity

(b) It is both antitussive and analgesic

(c) Only levo isomer is potent antitussive agent

(d) Levo isomer lacks both analgesic and antitussive property

18. The true statement following use of local anaesthetics is: -

    (a) Somatomotor nerve fibers are affected last

    (b) Autonomic fibers are affected first

    (c) Loss of motor power loss of sensation

    (d) Both a & b

19. The values of simple correlation co-efficient lie between

    (a) -a and a **(b) -1 and 1**

    (c) -0.9 and 0.9 (d) None

20. It is the value of the middle item of a given series of data arranged in ascending order of magnitude

    (a) Arithmetic mean (b) Mode

    (c) Median (d) None

21. Maximum permitted level of *E. coli* in frozen fish meant for export is

    (a) 100/g **(b) 20/g**

    (c) 0 (d) None

22. Change of state from solid to vapour without melting is known as

    (a) Sublimation (b) Evaporation

    (c) Condensation (d) None

23. Which statement pertaining to vitamin D is incorrect?

    (a) Sunlight through glass window is ineffective in producing vitamin D

    (b) Sunlight is more potent in temperate than tropic in producing vitamin D

    (c) Antirachitic power of sunlight is more potent during summer than winter.

    (d) Sunlight is more potent at high altitude

24. Tick the wrong statement regarding B-value ____________

    (a) Cereal has low B-value

    (b) Protein sources have low B- value

    (c) Mineral sources have high B-value

    (d) Limestone has a very high B-value

25. Which of the following statement is not true in man?

    (a) Sperms of two types

    (b) Egg of 2 types male & female producing

    (c) All eggs a like

    (d) None of the above

26. In genetics population means-

    (a) Breeding group (b) Mendalian group

    (c) Individual group (d) None of these

27. Dehorning means;

    (a) Removal of hoof (b) Removal of skin

    (c) Removal of hairs **(d) Removal of horns**

28. Manifestation of depressed appetite in nematode infection is a result of increased secretion of

    (a) *Histamine* **(b) *Cholecystokinin***

    (c) *Pepsinogen* (d) *Hydrochloric acid*

29. The correlation between breeding values and phenotypic values is equal to the square root of the

    (a) Standard error (b) Coefficient of variation

    (c) Sampling variance **(d) Heritability (Narrow sense)**

30. Correlation between repeated measurements of the same individual is called

    (a) Heritability (b) Realized Heritability

    (c) Repeatability (d) Breeding value

31. When repeatability of a trait is high it means

    (a) Specific environmental variance is least

    (b) Specific environmental variance is large

    (c) General environmental variance is large

    (d) General environmental variance is least

32. How much weight should gain during gestation for pregnant bitches;

    (a) 15 to 25 % (b) 10 to 20 %

    (c) 5 to 15 % (d) None of the above

33. The correlation in the breeding values among full-sibs is
    (a) 1/4 **(b) 1/2**
    (c) 1/8 (d) None of these

34. Solvency Status Of A Wool Industry Indicates That The Business Of Firm;
    (a) Is Loss Incurring
    (b) Is Sound
    (c) Is Neither Profit Nor Loss Making
    (d) Needs Attention

35. In RRB's The Share Of Center And States Will Be -
    (a) 40:60 (b) 80:20
    (c) 50:50 (d) 75:25

36. Agriculture is derived from two _______ *ager and cultura.*
    (a) Latin (b) Greek
    (c) American (d) Indian

37. Most of the plants die at _______ °C.
    (a) 0 to 10 (b) 15 to 30
    (c) 45 to 55 (d) 0 to - 10

38. The critical control point in HACCP system means:
    (a) Any location (b) Any practice
    (c) Any process **(d) All of above**

39. Puddling operation required by ...
    (a) Groundnut **(b) Paddy**
    (c) Cotton (d) Wheat

40. One cotton bale weight is ...
    (a) 180 kg **(b) 170 kg**
    (c) 138 kg (d) 190 kg

41. Indian cumin seeds contain volatile oil known as ...
    (a) Nicotine (b) Aldehyde
    (c) Malic acid **(d) Cuminol**

42. An underground small portion of sugarcane stem is known as ..

(a) Lamina (b) Tuber

(c) Shoot **(d) Rootstock**

43. Tobacco raised with low nitrogen and harvested by priming method is known as ...

(a) Rue curing (b) Culcutti

(c) Hookah (d) Cigarette

44. Which of the following is grassy weed?

(a) *Amaranthus viridis* **(b) *Cynodon dactylon***

(c) *Cyperus rotundas* (d) *Chenopodium album*

45. Which of the following is mimicry weed in paddy?

(a) *Striga spp.* **(b) *Avena spp.***

(c) *Digera arvensis* (d) *Cyperus spp.*

46. Difficult perennial weeds are also called ___________ weeds.

(a) Noxious (b) Objectionable

(c) Obnoxious **(d) Pernicious**

47. Which of the followings is acidophile weed?

(a) *Tribulus terrestris* (b) *Polygonum spp.*

(c) *Cynodon dactylon* **(d) *Rumex acetosella***

48. The negative pressure potential is often termed as

(a) Matric potential (b) Suction potential

(c) Pressure potential (d) None of these

49. A soil having 1000 cm^ volume and dry weight of 1.3 kg will have bulk density of.... g/cm^3

(a) 1300 (b) 130

(c) 1.3 (d) 13

50. The difference in soil moisture content between field capacity and the soil moisture content in the root zone before application of irrigation water is called ...

(a) NIR (b) GIR

(c) WR (d) None of these

Model Test Paper 4

1. The basic principle of taking crop rotation is ..,
   (a) To get higher crop production
   (b) To get higher return per unit area of the land
   (c) To maintain the soil fertility status of the soil
   (d) To keeps the weeds under control
2. Cropping intensity is ...
   (a) The extent of the use of land for purposes other than cropping
   (b) The extent of the use of land for cropping purposes during a given year
   (c) Net cropped area
   (d) Gross cropped area
3. Intercropping of sorghum with redgram is an example of...
   (a) Allelopathy (b) Annidation in space
   (c) Annidation in time (d) None of these
4. Cropping intensity means ...
   (a) Percentage ratio of gross cropped area to net cropped area
   (b) Percentage ratio of net cropped area to gross cropped area
   (c) Percentage ratio of the number of crops in rotation to period of one rotation
   (d) None of these
5. Blood mean contain ________ nitrogen.
   (a) 1-2 (b) 8-10
   (c) 13-20 (d) 5-8
6. Organic farming largely avoids or excludes the use of
   (a) Manures (b) Green manures
   (c) Biofertilizers additives **(d) Livestock feed**
7. The crops, in which the amount of cross-pollination is more than 5 % but the self-pollination is a rule, are considered as
   (a) Normally self-pollinated (b) Normally cross-pollinated
   (c) Often self-pollinated (d) Often cross-pollinated

8. It is often self-pollinated crop.
   (a) Wheat (b) Maize
   (c) Pearlmillet **(d) Rice**

9. The crops, in which the amount of self-pollination is more than 5 % but the cross-pollination is a rule, are considered as
   (a) Normally self-pollinated (b) Normally cross-pollinated
   (c) Often self-pollinated **(d) Often cross-pollinated**

10. It is often cross-pollinated crop.
   (a) Wheat (b) Maize
   (c) Pearlmillet (d) Rice

11. Isolation requirement for certified seed production of hybrid bajra is ...
   (a) 100 m **(b) 200 m**
   (c) 150 m (d) 300 m

12. In cases where wheat variety is susceptible to diseases caused by *Ustilago* spp. (loose smut) an isolation distance of ________ between seed field and other fields of wheat is recommended.
   (a) 3 m (b) 30 m
   (c) 80 m **(d) 180 m**

13. __________ is useful in curing fevers, worms,dysentery, general weakness, excessive gas formation in stomach and high blood pressure.
   (a) Isabgol (b) Senna
   (c) Ashwagandha **(d) Kalmegh**

14. For seed production pods of senna are collected during the month of...
   (a) October-November **(b) February-March**
   (c) May-June (d) December-January

15. The safed musli crop matures in about ________ days under cultivation.
   (a) 90-100 (b) 100-120
   (c) 120-150 (d) 150-180

16. Removal of ________ will give 20 % more yield of safed musli.
   (a) Suckers (b) Leaves
   (c) Rowering stalk (d) Rowers

17. Its root is used as a tonic, laxative, demulcent and expectorant.

(a) Liquorice (b) Safed musli

(c) Ashwagandha (d) Sarpagandha

18. __________kg/ha dry roots can be harvested from the fully matured liquorice crop.

(a) 1000-1500 (b) 1500-2000

(c) 2000-2500 (d) 2500-3000

19. Scientific name of satavari is ...

(a) *Andrographis paniculata* (b) *Chlorophytum borvilianum*

(c) *Asparagus racemosus* (d) *Pelargonium graveolens*

20. The origin of vetiver is ...

(a) China (b) Africa

(c) India (d) Brazil

21. Scientific name of lemon grass is ...

(a) *Vetiveria zizanioides* **(b) *Cymbopogon flexuosus***

(c) *Kaucha mucuna* (d) *Cyperus iria*

22. Which soil structure is most desirable for agriculture?

(a) Platy (b) Blocly

(c) Crumb (d) Granular

23. The family of patchouli is ...

(a) Labiatae (b) Liliaceae

(c) Asteraceae (d) Lamiaceae

24. When soil water potential is at hydrostatic pressure greater than atmospheric pressure, the pressure potential is

(a) Unity **(b) Positive**

(c) Negative (d) Zero

25. According to life cycle classification of spices, which crop belong to biannual?

(a) Onion (b) Cardamom

(c) Cumin (d) Dillseed

26. Genetics is the study of the __________ of the characteristics of an organism.
    (a) Inheritance (b) Variability
    (c) Both (A) & (B) (d) Neither (A) nor (B)
27. Viruses are minute particles that can infect ________
    (a) Bacteria (b) Animals
    (c) Plants **(d) All of the above**
28. Which of the following contributes maximum to Nitrogen fixation?
    (a) Combustion (b) Lightening
    (c) Biological Nitrogen Fixation **(d) Industries**
29. Conversion of atmospheric nitrogen to NH_3 by microorganism is referred to as :
    (a) Denitrification
    (b) Biological Nitrogen Fixation
    (c) Immobilization
    (d) Ammonification
30. Compost is considered superior if the organic matter content is __________
    (a) 10-20% **(b) 30-60%**
    (c) 20-30% (d) 1-10%
31. The science of atmosphere is called ...
    (a) Hydrology **(b) Meteorology**
    (c) Climatology
    (d) Aeronomy
32. The range of minimum thermometer is ...
    (a) -35 to + 55°C **(b) -40 to + 50°C**
    (c) -55 to + 60°C (d) -25 to + 50°G
33. Evapotranspiration is the highest when water supply is ...
    (a) Limited **(b) Unlimited**
    (c) Scarce (d) Withdrawn
34. The reflected radiation is measured by ...
    (a) Pyrheliometer **(b) Albedometer**
    (c) Hygrometer (d) Barometer

35. The Sun synchronous satellites move from ________

(a) East to West **(b) West to East**

(c) North to South (d) South to North

36. Agro meteorological observations at a location are recorded daily at...

(a) GMT (b) 1ST

(c) LMT (d) PST

37. It is the lowest temperature to which the air can be cooled by evaporating water in to it.

(a) Dry bulb temp. **(b) Wet-bulb temp.**

(c) Dew point temp. (d) Minimum temp.

38. Mahuva cake contains __________ alkaloid.

(a) Nimbidin (b) Ricin

(c) Saponin (d) None of these

39. High nitrate content in drinking water causes __________ disease in infants.

(a) Anemia **(b) Methamoglobinemia**

(c) Pneumonia (d) None of these

40. Rock phosphate is applied in _________ type soil.

(a) Alkali **(b) Acid**

(c) Neutral (d) None of these

41. How much of muriate of potash is required to supply 50 kg K_20/ha

(a) 50 kg (b) 73.3 kg

(c) 83.3 kg (d) 125.3 kg

42. Angle between costal and apical margin is _________

(a) Humeral angle **(b) Apical or outer angle**

(c) Anal or tomus angle (d) None of the above

43. Crickets belongs to family __________

(a) Acrididae (b) Tettigonidae

(c) Gryllidae (d) Gryllotalpidae

44. Hind wings are wanting and represented by a pair of slender process known as Halters found in __________

(a) Beetles in coleoptera

(b) House fly

(c) Grasshopper, cockroach and mantid

(d) Bugs

45. Wings are thin, transparent with many veins known as membranous found in ____________

(a) Termite, dragonfly and antiion

(b) Thrips and tur plume moth

(c) Grasshopper

(d) Beetles

46. Authority of extraction of nematodes from soil and plant nematodes

(a) Oostenbrink (b) Dr. Ritzemabose

(c) Dr. D. J. Raski (d) Dr. Seinhorst

47. An observer watching a sailing ship at sea notes that the ship appears to be "sinking" as it moves away. Which statement best explains this observation?

(a) The surface of the ocean has depressions

(b) The earth has a curved surface

(c) The earth is rotating

(d) The earth is revolving

48. Nematode take position for feeding is

(a) Retraction **(b) Exploration**

(c) Injection (d) Ingestion

49. The hard resting structure produced by fungal pathogen are

(a) Conidia (b) Sporangiospores

(c) Sclerotia (d) Zoospores

50. SiO_2 content of acidic igneous rocks is more than _________ %

(a) 65 (b) 55

(c) 44 (d) None of these

Model Test Paper 5

1. The minute colloidal clay particle is technically called __________
   (a) Micelle (b) Ped
   (c) Colloid (d) None of these
2. Dead heart in the wheat central shoot is a caused by
   (a) Stem borer (b) Root borer
   (c) Root aphid (d) None of the above
3. The boll worm which covers the opening once it enters into the boll is
   (a) Pink boll worm (b) Spotted boll worm
   (c) American boll worm (d) None of the above
4. Sugarcane setts filled with soil is characteristic symptom of
   (a) Shoot borer **(b) Termites**
   (c) Scale insect (d) Wooly aphid
5. Yellowish patches and silky webbing is observed on leaves due to
   (a) *Tetranychus urticae* **(b) *Tetranychus indica***
   (c) *Helicoverpa armigera* (d) *Thrips tabaci*
6. The cabbage caterpillar is ___________
   (a) Moth **(b) Butterfly**
   (c) Fly (d) None of these
7. Onion thrips shows _____________ appearance to the whole crop
   (a) Black (b) Blue
   (c) White (d) Yellow
8. Okra shoot and fruit borer ___________
   (a) *Helicoverpa annigera* **(b) *Earias vittella***
   (c) *Spodoptera litura* (d) *Aphis gossypii*
9. Forewing of the adult moth is pale with a wedge shaped green band in the middle..
   (a) Okra fruit borer (b) Brinjal fruit and shoot borer
   (c) Fruit borer (d) None of the above
10. The type of food chain in mangrove ecosystem is _________
    (a) Grazing food chain **(b) Detritus food chain**
    (c) Both (A) & (B) (d) None of the above

11. Byssinosis is an Occupational disease found in workers of
   (a) Coal mine **(b) Textile Industry**
   (c) Slaughter house (d) Galvanizing Industries
12. Diffusion will be at maximum rate in __________ state.
   (a) Solid (b) Liquid
   (c) Gas (d) Both (A) and (C)
13. A milk like preparation can be made from the seeds of
   (a) Gram (b) Grapes
   (c) Soybean (d) Barley
14. To absorb water and mineral salts
   (a) Root hair (b) Root cap
   (c) Side root (d) Tap root
15. Which of the following rules state that "in stable coordination structure, the total strength of valency bonds which reach an anion from all neighbouring cations equals the charge of anion"?
   (a) Pauling's first rule **(b) Pauling's second rule**
   (c) Pauling's third rule (d) None
16. An example of positive photoblastic seed is
   (a) Carrot **(b) Lettuce**
   (c) Silene (d) All of the above
17. Coconut milk factor is: .
   (a) An auxin (b) A gibberellined
   (c) Abscisic acid **(d) Cytokinin**
18. The height of troposphere is ________
   (a) 20 (b) 25
   (c) 30 (d) 35
19. Pulmonary oedema is related to _________ pollution
   (a) CO (b) NO_2
   (c) O_3 (d) SO_2
20. __________ is protect the earth from harmful UV.
   (a) O_2 (b) CO_2
   (c) O_3 (d) All of these

21. Expression of both alleles in F_1 is known as

    (a) Codominance (b) Incomplete dominance

    (c) Complete dominance (d) None of these

22. The gene having masking effect is called

    (a) Modifier gene (b) Enhancer gene

    (c) Epistatic gene **(d) Hypostatic gene**

23. Chemicals similar to DNA bases are called

    (a) Chemical DNA (b) Alkylating mutagen

    (c) Base analogs (d) EMS

24. The individuals with 2n + 1 chromosomes are commonly known as

    (a) Euploid **(b) Aneuploid**

    (c) Heteroploids (d) None of these

25. DNA as the genetic material was first discovered by

    (a) Griffith (1928)

    (b) Hershey and Chase (1951)

    (c) Avery, MacLeod & McCarty (1944)

    (d) Benzer (1955)

26. For plant height character parental genotypes AA = 20 cm and aa = 10 cm and hybrid Aa = 20 cm height.

    (a) Complete dominance (b) Incomplete dominance

    (c) Over dominance (d) Co-dominance

27. The common form of DNA present in living organisms is the

    (a) A form (b) Z form

    (c) C form **(d) B form**

28. A maintainer line is:

    (a) Male fertile and can restore fertility

    (b) Male sterile but cannot restore fertility

    (c) Male sterile but can restore fertility

    (d) Male fertile but cannot restore fertility

29. ICRISAT deals with:
    (a) Cotton (b) Maize
    (c) Millets (d) Wheat
30. Female flowers in cucumber can be increased by
    (a) Auxin (b) Ethylene
    (c) Both (A) & (B) (d) None of the above
31. Arka lalima and Arka kirtiman are the varieties of which crop
    (a) Potato (b) Cucumber
    (c) Tomato **(d) Onion**
32. Which cross between Totapuri x Kesar is free from malformation?
    (a) Sonpari (b) Sindhu
    (c) Pusa Arunima **(d) Saisugartha**
33. Which of the following is monoembryonic?
    (a) Sweet orange (b) Mandarin
    (c) Citron **(d) Pummelo**
34. The main principle of surveying is to work from
    (a) Part to the whole **(b) Whole to the part**
    (c) Higher to lower level (d) Lower to higher level
35. It is the rate of doing work
    (a) Work done (b) Torque
    (c) Power (d) Energy
36. In hydraulic sprayers, the degree of atomisation is primarily a function of
    (a) Liquid pressure and the nozzle characteristics
    (b) Air velocity
    (c) Size of the nozzle
    (d) Size and shape of the atomizer
37. Planter differ from seed drill in respect of
    (a) Kind of power transmission
    (b) Kind of metering mechanism
    (c) Kind of furrow opener used
    (d) All of the above

38. Sweeps are used for
    (a) Seedbed preparation (b) Ridging
    (c) Earthing plants **(d) Mulching**

39. The main disadvantage of wind power is that
    (a) It is available only in coastal areas
    (b) Wind energy systems are noisy when in operation
    (c) Large land area is required
    (d) The capacity utilization is less

40. The gas produced by burning wood in an insufficient supply of oxygen is called
    (a) Producer gas (b) Biogas
    (c) Natural Gas (d) Nitrogen gas

41. The specific heat of dry grains in Kcal/kgo C range from:
    (a) 0.25-0.35 **(b) 0.35-0.45**
    (c) 0.40-0.60 (d) 0.80-1.0

42. Natural air drying needs dehumidification if the atmospheric humidity is:
    (a) Low (b) Intermediate
    (c) High (d) Very high

43. Ideal temperature for preservation of semen in liquid nitrogen
    (a) -150°C **(b) -196°C**
    (c) -159°C (d) -169°C

44. The head quarter of GDDC is located at
    (a) Gandhinagar (b) Anand
    (c) Mahesana (d) Delhi

45. ____________ is known as much more than half of the herd
    (a) Cattle (b) Buffalo
    (c) Bull (d) Bullock

46. __________ refers to shift in consumption behaviour of an individual to other related goods due to rise in price of a particular good.
    (a) Income effect **(b) Substitution effect**
    (c) Price effect (d) None of these

47. _________ refers to various quantities of goods offered by the seller at various price.

(a) **Supply** (b) Demand

(c) Both (A) & (B) (d) None of these

48. Tax bearer and tax payer are different in case of

(a) Direct Tax (b) **Indirect tax**

(c) Both (A) & (B) (d) None of these

49. There will be loss with the rise in general price level to ___________

(a) **Creditor** (b) Debtor

(c) Both (d) None of these

50. If money circulation increases the prices in economy ___________

(a) **Increases** (b) Decreases

(c) Constant (d) None of these

Model Test Paper 6

1. Regularity market committee consisting of __________ member

   (a) 17 (b) 27

   (c) 19 (d) 7

2. Which of the following is the leader in the field of cooperative marketing within the country as well as in the export markets?

   (a) NCDC **(b) NAFED**

   (c) AGMARK (d) FCI

3. CCIS was introduced by GCI in all the states by the year _________

   (a) 1980 (b) 1975

   (c) 1985 (d) 1990

4. All India Rural Credit Survey Committee appointed by the RBI in _____________ under the chairmanship of Shri A.D. Gorwala.

   (a) 1951 (b) 1971

   (c) 1961 (d) 1981

5. _________ serve as a link between the RBI and PACS.

   (a) SCBS (b) CLDB

   (c) DCB (d) PLDB

6. Forecasting describes what one ________ to happen if no changes are made to escape that happening.

   (a) Wants (b) Goods

   (c) Expects (d) None of these

7. Additional cost incurred to produce additional unit of output is

   (a) Marginal cost (b) Marginal input cost

   (c) Average cost (d) Opportunity cost

8. Which of the following curves cuts the AC curve at its lowest point?

   (a) AVC curve (b) AFC curve

   (c) Opportunity cost curve **(d) MC Curve**

9. In production function, when MPP zero,TPP

   (a) Increases at increasing rate (b) Increases at decreasing rate

   (c) Maximum (d) Zero

10. The speed of CPU measured in

(a) Mhz (b) Bytes

(c) Sec (d) Kbps

11. Incase of normal distribution, the values of all odd moments is equal to :

(a) Zero (b) One

(c) Infinity (d) None of these

12. Replication in an experiment means:

(a) The number of times a treatment repeated in an experiment

(b) The numbers of blocks

(c) The total numbers of treatments

(d) None of these

13. From point estimation, we always get:

(a) Single value (b) Two values

(c) Range of values (d) Zero

14. Where is the headquarter situated for AICRP on Wheat & Barley Improvement Project

(a) Karnal (b) Hyderabad

(c) Jodhpur (d) Bangalore

15. A polling agency conducted a survey of 100 doctors on the question "Are you willing to treat women patients with the recently approved pill RU-486"? The conservative margin of error associated with the 95% confidence interval for the percent who say 'yes' is

(a) 50% **(b) 10%**

(c) 5% (d) 2%

16. Where is the headquarter situated for AICRP on Small Millets,

(a) Karnal (b) Hyderabad

(c) Jodhpur **(d) Bangalore**

17. xtension when aims at empowering or uplifting the poor:

(a) Emancipatory extension (b) Formative extension

(c) Informative extension (d) Persuasive extension

18. IVLP stands for:
    (a) Indian Village Linkage Programme
    (b) Indian Village Linkage Plan
    (c) Institution Village Linkage Programme
    (d) Integrated Village Lab to Land Programme
19. The number of V.L.W.'s is an I.A.D.P. block is :
    (a) 5 (b) 10
    (c) 15 **(d) 20**
20. Where is the headquarter situated for AICRP Tropical Fruits
    (a) Jhansi **(b) Bangalore**
    (c) Lucknow (d) Bikaner
21. Where is the headquarter situated for AICRP Cashew
    (a) Varanasi (b) Shimla
    (c) Thiruvananthapuram **(d) Puttur**
22. Since 1950-51 India through its green revolution technology increased production of eggs by
    (a) 6-fold **(b) 27-fold**
    (c) 5-fold (d) 15-fold
23. The relatively non-mobile among the TOT programmes of ICAR:
    (a) LLP (b) ORP
    (c) AICPND **(d) KVK**
24. State with highest percentage of area under acidic soil
    (a) Uttar Pradesh **(b) Nagaland**
    (c) Kerala (d) Punjab
25. MSW abbreviates for which of the following
    (a) High-density Polyethylene
    (b) Hazardous Waste Management
    (c) Maximum Contaminant Level
    (d) Municipal Solid Waste
26. SS abbreviates for which of the following
    (a) Reinforced concrete pipe (b) Soil conservation service
    (c) Suspended solids (d) Sewage treatment plant

27. TOC abbreviates for which of the following
    (a) Total dissolved solids (b) **Total organic carbon**
    (c) Total solid (d) Total suspended solids
28. The light emitted from the sun reaches the earth in
    (a) 8.3 minutes (b) 18.3 minutes
    (c) 28.3 minutes (d) 38.3 minutes
29. It is used for measuring beam radiation.
    (a) Pyranometer **(b) Pyrheliometer**
    (c) Both (d) None
30. Which of the following weed is as commonly known as the Bengal dayflower or tropical spiderwort, field bind weed.
    (a) Sphenoclea zeylanica (b) *Marsilea quadrifolia*
    (c) *Commelina benghalensis* (d) *Echinochloa oryzoides*
31. The attainment of perfection demands that undesirable and useless movements are replaced by desirable and useful movements:
    (a) Principle of motivation (b) Principle of attitude
    (c) Principle of readiness (d) Principle of satisfaction
32. Annually, India is losing nearly of nitrogen,
    (a) 0.8 million tones (b) 1.8 million tonnes
    (c) 26.3 million tonnes (d) None
33. Which one is a non projected aid
    (a) Cinema (b) Slide
    (c) Over head projector **(d) Picture**
34. Gurgaon Project started in the year
    (a) In 1903 **(b) In 1920**
    (c) In 1928 (d) In 1932
35. The movie film recording is made by the _____ process
    (a) Mechanical (b) Magnetic
    (c) Optical (d) None of these
36. National Dairy Research Institute
    (a) Shimla (b) Solan
    (c) Karnal (d) Hissar

37. National Bureau of Animal Genetic Resources

(a) Shimla (b) Solan

(c) Karnal (d) Hissar

38. The process of selecting information that is relevant to a problem while ignoring distractions

(a) Selective encoding (b) Selective decoding

(c) Selective comparison (d) None of these

39. How old is Shifting Agriculture

(a) Neolithic (about 7,000 B.C.) (b) 500–1450 A.D

(c) 18th Century (d) 20th Century

40. Medieval Agriculture

(a) Neolithic (about 7,000 B.C.) (b) 500–1450 A.D

(c) 18th Century (d) 20th Century

41. Earliest historical record of Dogs were domesticated in Iraq

(a) 70 million years ago (b) 40 million years ago

(c) 10 million years ago (d) 8700 B.C.

42. Charles Darwin was the first to

(a) Discovered the laws of heredity.

(b) Experiments on cross and self-fertilization in plants.

(c) Proposed "Malthusian Theory"

(d) Theory of "Optima and Limiting Factors"

43. Autumn rice crop ' Virippu ' is known in which parts of India

(a) West Bengal (b) Assam

(c) Orissa **(d) Kerala**

44. Winter rice is known as 'Agahani' names in which part of India

(a) West Bengal (b) Assam

(c) Orissa **(d) Bihar and Uttar Pradesh**

45. The stage of adoption wherein group discussion is most effective

(a) Awareness **(b) Evaluation**

(c) Testing (d) Persuasion

46. Long term preservation of seeds through cryogenics by using

    (a) Liquid nitrogen (b) Controlled conditions

    (c) Cold storage (d) None

47. Action-reaction interdependence in communication is called

    (a) Credibility **(b) Feed back**

    (c) Message treatment (d) Fidelity

48. Which of the following scientist has given the theory of information of communication?

    (a) Shannon and Weaver (b) Leagon

    (c) Berio (d) Aristotle

49. It is feelings of an individual towards or against something.

    (a) Skill (b) Knowledge

    (c) Attitude (d) Motivation

50. Which of the following vitamin cures Melanea

    (a) A (b) B

    (c) C (d) D

Model Test Paper 7

1. Agglutination tests is done for
   (a) Tests to identify bacteria
   (b) Tests used to detect Brucellosis in cattle
   (c) Both
   (d) None
2. Algae grow rapidly in water which is rich in
   (a) Phosphates (b) Nitrates
   (c) Sulphates (d) All
3. A chemical substance which kills trees
   (a) Arboricide (b) Acaricide
   (c) Pesticide (d) Herbicide
4. Improvement in mean genotypic value of the selected families over base population is known as
   (a) Genetic advance (b) Genetic load
   (c) Heritability (d) None
5. When excess of ionic concentration is formed near the root surface then back movement of ions takes to soil solution.
   (a) Back diffusion (b) Back furrow
   (c) Background region (d) Background
6. A computer program that allows the user to determine the positions of atoms or ions in biological molecules.
   (a) Biogenesis (b) Biogeochemistry
   (c) Biograf (d) Biological availability
7. The total mass of living organisms in a given volume or mass of soil.
   (a) Biological denitrification (b) Biological N fixation
   (c) Biomagnification **(d) Biomass**
8. A ferruginous deposit in bogs and swamps formed by oxidizing algae, bacteria or the atmosphere on iron in solution.
   (a) Blue vitriol (b) Blue-green algae
   (c) Bog **(d) Bog Iron Ore**

9. The temperature at which the vapor pressure of a liquid is equal to the applied pressure; also the condensation point

   (a) Boiling Point (b) Bolting or Shooting

   (c) Bolus (d) Boliland rice

10. Pair of electrons involved in a covalent bond.

   (a) Bond energy (b) Bond Energy

   (c) Bond length **(d) Bonding Pair**

11. The rate of change in plant material (dry weight) per unit of time.

   (a) Absolute growth rate (AGR) **(b) Absolute humidity**

   (c) Absolute temperature (d) Absolute water requirement

12. Movement of ions and water into the plant root as a result of metabolic processes by the root against an activity gradient.

   (a) Absorption spectrum **(b) Absorption, active**

   (c) Absorption, passive (d) Absorptivity

13. A reaction in which two atoms or groups of atoms are added to a molecule, one on each side of a double or triple bond. A reaction where atoms add to a carbon-carbon multiple bond.

   (a) Adaptation (b) Adaptive Radiation

   (c) Addition compound **(d) Addition Reaction**

14. Collection of various organic and inorganic substances in soil that are capable of adsorbing ions and molecules.

   (a) Adsali sugarcane (b) Adsorbate

   (c) Adsorption **(d) Adsorption complex**

15. Rice that is kept at least 4 months after harvest. Expands more on cooking and less sticky than cooked, freshly harvested rice.

   (a) Agar (b) Agarose

   (c) Agate **(d) Aged rice**

16. These are flooded soils where hydrogen sulphide is formed due to sulphate reduction and anaerobic decomposition of organic matters.

   (a) Akinete **(b) Akiochi soils**

   (c) Alabaster (d) Alban

17. An extrusive igneous rock that develops from a magma that is chemically between felsic and mafic and whose mineral crystals are fine.

(a) Anatose (b) Andalusite -$Al_2 SiO_5$.

(c) Andesite (d) Angelite

18. A measure of how efficient an emission process is.

(a) Quantity intensity ratio (b) Quantum mechanics

(c) Quantum Yield (d) Quantum

19. Mineral with the chemical formula SiO_2.

(a) Quartering **(b) Quartz**

(c) Quartzite (d) Quaternary

20. Potential users of Knowledge Management Portal includes

(a) Farmers (b) Scientists

(c) Extension officials **(d) All**

21. Which are important pillars of knowledge management

(a) Information (b) Data management

(c) All (d) None

22. Warana Wired Village Project, Warananagar main focus is on

(a) Transactions with cooperative (b) Cropping knowledge

(c) Basic computer literacy **(d) All**

23. Indian legislation of Plant Variety Protection and Farmers Rights include

(a) Plant Breeders' Rights (b) Researcher exemption

(c) Farmers' rights **(d) All of the above**

24. Which of the following nitrifying bacteria form 'Cells in masses in slime without a common membrane'

(a) Nitrosomonas (b) Nitrosococcus

(c) Nitrosogloea (d) Nitrosospira

25. The tilt angle in the disc plough

(a) 15-25 (b) 35-40

(c) 10-12 (d) 40-50

26. The restriction enzyme EcoRI generates which type of DNA fragments

(a) 3' sticky ends **(b) 5' sticky ends**

(c) 3' blunt ends (d) 5' blunt ends

27. Plasmid vectors can be designed with which of the following features
    (a) Antibiotic resistance
    (b) Colorimetric "markers"
    (c) Strong or weak promoters for driving expression of a protein
    (d) All of the above
28. What are the problems with conventional cloning
    (a) Inconvenient restriction sites
    (b) Vector construction is laborious
    (c) Time-consuming reactions
    (d) All of the above
29. Insecticidal activity of the inclusion bodies was demonstrated for the first time by
    (a) Thomas Angus (b) Philip Fitz-James
    (c) Ishiwata (d) Berliner
30. *Bacillus thuringiensis sub* sp israelensis is toxic to
    (a) Termite (b) Aphids
    (c) Mites **(d) Housefly**
31. Which of the following serves as a termination codon
    (a) UAA (b) UAG
    (c) UGA **(d) All of the above**
32. The original source of Ti plasmids is
    (a) Agrobacterium (b) Bacillus thuringiensis
    (c) Arthrobacter (d) Pseudomonas
33. The advantages of ISSR (Inter Simple Sequence Repeat) are
    (a) Discriminative (b) Reproducible
    (c) Simple protocol **(d) All of the above**
34. What is the reason behind presence of great diversity in Bt Genes?
    (a) Presence of Cry genes on megaplasmids
    (b) Differences in physical environment
    (c) Vertical gene transfer
    (d) None of the above

35. Sources of B1
    (a) Whole grains (b) Dried legumes
    (c) Liver **(d) All**
36. Disaccharides (composed of two monosaccharides) example
    (a) Glucose (b) Sucrose
    (c) Lactose **(d) Both b & c**
37. Which of the following is Nonexpanding Type clay minerals
    (a) Illite (b) Chlorites
    (c) Both (d) None
38. What is the multiplying factor for converting N into NH3
    (a) 0.36383 (b) 0.77446
    (c) 1.2159 (d) 1.2884
39. ADCO process a system of composting was initiated by
    (a) Hutchinson, H.B and Richards, E.H.
    (b) Fowler and Ridge
    (c) Howard and Ward
    (d) Dr. C.N.Acharya
40. Most soils have a water content
    (a) Less than porosity (b) More than porosity
    (c) Same asporosity (d) None
41. Typical water content (vol/vol) in Saturated water content
    (a) 0 0.2–0.5 (b) 0.1–0.35
    (c) 0.01–0.25 (d) 0.001–0.1
42. Suction pressure(J/kg or kPa in Saturated water content
    (a) 0 (b) –33
    (c) –1500 (d) Infinity
43. Mesocarp is related to
    (a) Bast fibre (b) Life belts
    (c) Old world cotton **(d) Cocus nucifera**
44. Form bacteria is related to
    (a) Non-proteinaceous **(b) Nucleoid**
    (c) Wall-less bacteria (d) Gram positive

45. Notable year for Ever-Green Revolution

(a) 1968 **(b) 1996**

(c) 2005 (d) 2010

46. Three Largest Producing States in decreasing order of production of Rice during 2008-09

(a) West Bengal-Andhra Pradesh-Uttar Pradesh

(b) Uttar Pradesh-Punjab-Haryana

(c) Andhra Pradesh-Karnataka-Rajasthan

(d) Rajasthan-Karnataka-Maharashtra

47. Average concentration in plant tissue of Chlorine, Cl

(a) 0.1% **(b) 100mg/kg**

(c) 100mg/kg (d) 20mg/kg

48. 1 ton equals

(a) 2000 lbs (b) 200 lbs

(c) 20 lbs (d) 2 lbs

49. What is the factor for converting Cubic centimeters into Cubic inches

(a) 6.102×10^{-2} (b) 1×10^{-6}

(c) 3.53×10^{-5} (d) 2.642×10^{-4}

50. Indian Journal of Agricultural economics is published from

(a) Mumbai (b) Chennai

(c) New Delhi (d) Dharwad

Model Test Paper 8

1. Jean Derze has worked on this aspect of Indian Economy
   (a) Development Economics (b) Monetary policy
   (c) Industrial Economics (d) Public Finanace
2. Tractor and Farm Equipment (TAFE) is a unit of
   (a) Amalgamation Group (b) Murugappa Group
   (c) Mahindra and Mahindra Group (d) Tata Group
3. Which of the following measures is most necessary for the use of modern farm implementation?
   (a) Marketing facilities (b) Irrigation facilities
   (c) Cooperative facilities **(d) Consolidation of holdings**
4. The chairman of 1900 Famine commission was
   (a) A.P Macdonnel (b) A, R.Desai
   (c) M. Sen (d) Powell
5. The population obstacle to Econoic Betterrment wasthe work of
   (a) Spenger (b) Hasan ali
   (c) A.G. Noorani (d) Lewis
6. The Economic History of British India was written by
   (a) R.B. Saha (b) Gadgill
   (c) R.C. Dutt (d) M.G.Ranade
7. He is a creative individual and likely to be an entrepreneur
   (a) Reformist (b) Retualist.
   (c) Retreatist **(d) Innovator**
8. The French word 'Entreprendre' means:
   (a) To do something (b) To organize
   (c) To innovate (d) To take risk
9. The persons who prefer to continue in the same venture by transforming themselves to fit the changing demands is/are known as:
   (a) Managerial Entrepreneur (b) Mobile Entrepreneur
   (c) Innovative Entrepreneur (d) Empire builders

10. Which of the following book is written by Randhava G.S. and M.A. Mukhopadhyay.?
    (a) Introduction to floriculture **(b) Floriculture in India**
    (c) Commercial flowers (d) Advantage in horticulture
11. Which of the following is not a Rosa species grown for oil extraction?
    (a) *Indica* (b) *Damascene*
    (c) *Bourboniana* (d) *Centifolia*
12. Water disinfection is done by ______
    (a) Sodium chloride **(b) Sodium hypochloride**
    (c) Potassium chloride (d) Tween 20
13. Vegetables are valued for their high _________
    (a) Vitamins (b) Starch
    (c) Fat (d) Protein
14. Commercially gum is extracted from
    (a) Cowpea (b) Tomato
    (c) Pea **(d) Cluster bean**
15. Which is most serious diseases of tomato?
    (a) YVMV (b) BFSB
    (c) TSWV (d) None of these
16. NPK requirement of sponge gourd is?
    (a) 50 : 25 : 25 (b) 25 : 50 : 50
    (c) 25 : 50 : 25 (d) 25 : 25 : 50
17. The crop grown for green manuring
    (a) Chilli (b) Brinjal
    (c) Sunhemp (d) Okra
18. The practice of growing arable crops between two subsequent rows of leguminous shrubs is called as……
    (a) Lay farming (b) Alley farming
    (c) Alley cropping (d) Alternate land use
19. Parboiling operation is followed in……. crop.
    (a) Rice (b) Groundnut
    (c) Cotton (d) Sugarcane

20. Vertical mulch is used in soils ....

    (a) Black cotton soil (b) Red soil

    (c) Laterite soil (d) All of these

21. _____________ is the father of modern enzymology.

    (a) Rose (b) knoop

    (c) J.B Sumner (d) Bloor

22. % K = % K2 O * __________

    (a) 0.43 **(b) 0.83**

    (c) 1.33 (d) 2.24

23. C:N ratio of the Wheat straw is ________________________

    (a) 400:1 (b) 100:1

    (c) 80:1 (d) 20:1

24. According to elemental composition of undecomposed organic matter contain ________________ % of ash

    (a) 40% (b) 10%

    (c) 8% (d) 25%

25. Recently formed soils are called

    (a) Entisols (b) Aridisols

    (c) Inceptisols (d) Mollisols

26. Which are the chloride loving plants

    (a) Grape **(b) Oil palm and coconut**

    (c) Tobacco and Tomato (d) Onion

27. The process of deposition of soil material in the lower layer is called

    (a) Eluviation **(b) Illuviation**

    (c) Podzalization (d) Pedoturbation

28. Mechanization possibility is strongly influenced by

    (a) Farm size (b) Cost of farm power

    (c) Availability of suitable machines **(d) All of the above**

29. The material used for cutting blade of rotavator is

    (a) Carbon steel **(b) High carbon steel**

    (c) Low carbon steel (d) Medium carbon steel

30. Tractor manufacturing was started in India in the year

(a) 1951 **(b) 1961**

(c) 1971 (d) 1981

31. The improved kothar type structure generally made of

(a) 5 cm thick (b) 10 cm thick

(c) 15 cm thick (d) 20 cm thick

32. The phenomenon of the reversion of mature cells to the meristematic state leading to the formation of callus is known as

(a) Redifferentiation **(b) Dedifferentiation**

(c) Organogenesis (d) Embryogenesis

33. Tissue plasminogen activator is

(a) A vitamin

(b) A chemical that stimulates tissue differentiation

(c) An enzyme

(d) An electric device

34. One of the following would be called VNTR

(a) TTTTCCCC **(b) GTGTGTGT**

(c) GGTTGGTT (d) GGGGTTTT

35. Essential constituents of a cloning vector are

(a) Replicon, promoter and selectable marker

(b) Replicon, unique restriction site and selectable marker

(c) Promoter, operatop and restriction site

(d) Replicon, unique restriction site and promoter

36. When producing human serum albumin (HSA) in transgenic plants, what would be an advantage of incorporating the transgene into the chloroplast genome?

(a) HSA diverts too many amino acids if expressed in cytoplasm

(b) HSA becomes allergenic if expressed in cytoplasm

(c) HSA is toxic to plants if expressed in cytoplasm

(d) HSA is degraded by plant enzymes if expressed in cytoplasm

37. Pusa Ratna is a new variety of

    (a) Wheat (b) Soybean

    (c) Rice **(d) Mungbean**

38. Dr. Har Govind Khurana was awarded the Nobel Prize in 1968 for his work on

    (a) Mutation

    (b) Amino acids synthesis

    (c) Genetic code and in vitro synthesis of DNA

    (d) RNA synthesis

39. The number of single crosses will be equal to

    (a) n(n-1)/2 (b) n(n-1) (n-2)/8

    (c) n(n-1) (n-2) (n-3)/8 (d) n(n+1) (n+2)/4

40. Which can function as carrier in active absorption?

    (a) Cytochrome (b) Lecithin

    (c) Ferredoxin (d) Plastoquinone

41. Major role of minor elements inside living organisms is to act as

    (a) Binder of cell structure

    (b) Constituent of hormones

    (c) Building blocks of important amino acids

    (d) Cofactor of enzyme

42. Sink is related to

    (a) Phytochrome (b) Enzymes

    (c) Stomata **(d) Transport of minerals**

43. At high oxygen concentration, the rate of photosynthesis decreases due to

    (a) Warburg effect (b) Pasteur effect

    (c) Emerson effect (d) Richmond Lang effect

44. Photosynthetic enhancement with flashing light was first observed by

    (a) Benson and Calvin (b) Hill and Calvin

    (c) Hatch and Slack **(d) Emerson and Arnold**

45. Sandal spike disease of wheat is caused by...........

(a) Fungi (b) Bacteria

(c) Virus (d) MLO

46. Consider the following processes on host plant occurring during pathogenesis ________

1. Landing of inoculum
2. Penetration
3. Germinatiom
4. Recognition
5. Establishment and sporulation

The correct sequence of these processes is

(a) 1,2,3,4,5 (b) 2,3,1,5,4

(c) 1,3,2,4,5 (d) 4,1,2,3,5

47. Which one of the following is odd

(a) Ascospores (b) Cygospores

(c) Clamydospores **(d) Oospores**

48. An example(s) of a relative method to assess economic threshold levels for an insect pest is:

(a) Number of insects per leaf (b) Number of insects per plant

(c) Number of insects per twig **(d) A, B and C**

49. Examples of some density-independent mortality factors in insect populations are:

(a) Predators, parasitoids, and pathogens.

(b) Predators, chemical insecticides, and intraspecific competition.

(c) Chemical insecticides, hurricanes, and temperature extremes (i.e., hot and cold)

(d) Chemical insecticides, flooding, and intraspecific competition.

50. Which of the following responses occur when predators interact with prey populations?

(a) A numerical response (b) A functional response

(c) A lag effect **(d) a and b only**

Model Test Paper 9

1. To which of these groups do insect pathogens belong?
   (a) Viruses and bacteria (b) Protozoa and fungi
   (c) Bacteria and protozoa **(d) All of these**
2. To which body segment are the halters attached?
   (a) Mesothorax (b) Prothorax
   (c) First abdominal **(d) Metathorax**
3. Which part of the exoskeleton lies between the wax layer and the cement layer?
   (a) Exocuticle (b) Cuticulin layer
   (c) Endocuticle **(d) None of these**
4. The cibarium is best described as:
   (a) Thoracic muscles that move the wings
   (b) A structure on the pretarsus
   (c) The innermost layer of the epicuticle
   (d) A muscular pump that sucks food into the mouth
5. Fly larvae (maggots) move away from a bright source of light. This is an example of a
   (a) Taxis (b) Reflex
   (c) Kinesis (d) Transverse orientation
6. Hollowcone nozzle is used to spray
   (a) Weedicide **(b) Insecticides**
   (c) None of them (d) All
7. Most of the caterpillar have legs or abdominal legs on segments
   (a) 2-5 & 10 **(b) 3-6 & 10**
   (c) 6-10 (d) 5-8 & 10
8. Trade name of Difenthioron is
   (a) Endosulfon (b) Carbofuron
   (c) Polo (d) Bestox
9. Insects feeding on plants of several genera within a family are called
   (a) Phytopagous (b) Polyphagous
   (c) Oligophagous (d) Monophagous

10. Which of the following order comprises parasitic insects
    (a) Hymenoptera (b) Thysanoptera
    (c) Phasmida (d) Lepidoptera

11. Which of the following produces more lac, & is important for commercial production
    (a) Male **(b) Female**
    (c) Nymph (d) All of the above

12. The interaction in which at least one species is harmed is called as
    (a) Positive interaction **(b) Negative interaction**
    (c) Commensalisms (d) None of these

13. Viviparous insects produce
    (a) Eggs **(b) Young ones**
    (c) Nymph (d) None of above

14. A substance that is destructive to nerve tissue is called
    (a) Toxicant **(b) Seuath**
    (c) Aerosol (d) Toxin

15. The concentration of Juvenile Hormone is high at
    (a) Egg stage (b) Pupal stage
    (c) Adult stage (d) All

16. Pupae having appendages free and visible externally is present in
    (a) Butterflies **(b) Beetles**
    (c) Flies (d) Moths

17. Which hormone cause moulting
    (a) Ecdysome (b) Ecdysis
    (c) Exuvium (d) Moulting

18. Method of Expressing moisture in grain:
    (a) Wet Weight basis (b) Dry weight basis
    (c) Both a and b (d) None of all

19. Under some circumstances a host may pass through the most susceptible stage quickly or at a time when insect numbers are reduced. Such type of psuedoresistance is known as:
    (a) Escape **(b) Host Evasion**
    (c) Susceptibility (d) All of them

20. Oligogenic resistance is also called ____________

    (a) Major gene resistance (b) Minor gene resistance

    (c) Both (a and b) (d) None of all

21. Poisoning symptoms of Rotenone.

    (a) Inactiveness (b) Restlessness

    (c) Fanning movement (d) Jitters

22. Nature of action of Chlorthion insecticide.

    (a) Chitin inhibitor (b) Fumigant

    (c) Contact (d) Systemic

23. In orthoptera, speialized _______ organs are present.

    (a) Auditory (b) Stridulatory

    (c) Auditory and stridulatory (d) None

24. Water bugs belong to the sub order ____________

    (a) Cryptocerata (b) Gymnocerata

    (c) Adephaga (d) None

25. Antlion belongs to the family ____________

    (a) Ascalaphidae **(b) Myrmeleontidae**

    (c) Nemopteridae (d) None

26. Seed renewal period for bajra variety is

    (a) 2 year (b) 1 year

    (c) 3 year (d) 6 months

27. Progeny of nucleus seed

    (a) Breeder seed (b) Foundation seed

    (c) Registered seed (d) Certified seed

28. Sucrose is a

    (a) Monosaccharide **(b) Oligosaccharide**

    (c) Trisaccharide (d) Polysaccharide

29. Endoplasmic reticulum is a

    (a) Cell inclusions (b) Living

    (c) Cell organelles **(d) Both b & c**

30. Precursor for ethylene synthesis is
    (a) Methionine and b-alanine (b) **Tryptaphene**
    (c) Phytiri (d) Glutaline
31. In Synopsis, the type of dormancy is
    (a) Morphological (b) Physiological
    (c) Physical (d) Both a & c
32. The color of amylase on exposure to iodine solution
    (a) Yellow (b) Green
    (c) Blue (d) Purple red
33. Albumins are coagulated by
    (a) Cooling (b) Pasturization
    (c) Heating (d) None
34. In seeds energy mobilization is enhanced by the enzyme
    (a) Lyases (b) Catalases
    (c) Aluerone layer (d) Embryo
35. Type of dormancy in *Terminalia* spp.
    (a) Mechanical dormancy (b) Physical dormancy
    (c) Physiological dormancy (d) Combined dormancy
36. Cleavation of internal peptide bonds is carried by enzymes
    (a) Endopeptidases (b) Ectopepciidases
    (c) Proteases (d) Amino peptidases
37. In lettuce, alpha galactosidase activity is promoted by
    (a) Cotyledon **(b) Embryonic Axis**
    (c) Testa (d) Endosperm
38. In seeds production of alpha amylase is
    (a) De-novo (b) Direct
    (c) Indirect (d) 1 & 2
39. Normally seed extraction is not applicable to
    (a) Pterocarpus (b) Conifers
    (c) Serotinous fruits **(d) Dipterocarpus**
40. Color grading could be adapted to separate good seeds in
    (a) *Acacia mellifera* (b) *Acacia nilotica*
    (c) *Cassia fistula* (d) *Cassia hybrida*

41. Wet extraction is the common practice of extraction in
    (a) Chillies (b) Bhendi
    (c) Onion **(d) Bitter gourd**
42. Hardened seed performed well in
    (a) Marginal lands (b) Wet lands
    (c) Rainfed sowing (d) Garden land
43. Karnal bunt of wheat can be identified by
    (a) NaOH soaking test (b) Blotter method
    (c) PCNB (d) Agar plate method
44. ISTA was established during
    (a) 1968 (b) 1966
    (c) 1924 (d) 1997
45. Peroxidase test for soybean is useful to assess
    (a) Viability **(b) Varietal purity**
    (c) Vigor (d) Mechanical damage
46. In certification, the number of counts to be taken from 6 to 10 acres
    (a) 6 (b) 7
    (c) 8 (d) 9
47. Decrease in demand means
    (a) Demand change due to change in price
    (b) Demand changes not due to the change in price but due to income
    (c) Both a & b
    (d) None of these
48. The maximum error degree of freedom can be provided for same numbers of treatment and replication of a experiment by:
    (a) RBD **(b) CRD**
    (c) LSD (d) None of these
49. Lab to Land programmed is associated with ICARs ....
    (a) Golden jubilee (b) Silver jubilee
    (c) Diamond jubilee (d) Platinum jubilee
50. Age is classified under ....
    (a) Discontinuous variable (b) Discrete variable
    (c) Continuous variable (d) Intervening variable

Model Test Paper 10

1. Modified stem of banana is called as ——————

   (a) Sword suckers **(b) Rhizome**

   (c) Water sucker (d) Corm

2. Among fruits the highest vitamin 'A' content is in mango and it is about__________

   (a) 200 IU/ 100 g (b) 1500 IU/ 100 g

   (c) 4000 IU/ 100 g (d) 300IU/ 100g

3. Pomology is defined as science of cultivation of—————

   (a) Fruits (b) Vegetables

   (c) Flowers (d) None of these

4. Acta horticulture is a publication of————————

   (a) ASHS, USA **(b) ISHS, Belgium**

   (c) JSHS, Japan (d) HIS, India

5. Kagzi lime is commercially propagated by ——————

   (a) Budding (b) Layering

   (c) Cutting **(d) Seed**

6. Best method of vegetative propagation in pomegranate is

   (a) Veneer grafting (b) Patch budding

   (c) Chip budding **(d) Air layering**

7. Grapevine will start fruiting after _________of planting.

   (a) 4 years (b) 5 years

   (c) 1.5 years (d) 2.5 years

8. Maximum number of fruit plants can be planted in the orchard by the system :

   (a) Digonal system (b) Square system

   (c) Hexagonal system (d) Rectangular system

9. Which of the following is not a auxin ?

   (a) GA_2 (b) NAA

   (c) 2,4 D (d) IAA

10. Which of the following not the kind of budding?
    (a) Patch (b) Shield
    (c) Ring **(d) All of above**
11. The cultivated strawberry having diploid chromosome number Zn = __________
    (a) 28 **(b) 14**
    (c) 30 (d) 20
12. Which year is celebrated as the United Nations Decade on Biodiversity?
    (a) 2001-2010 **(b) 2011-2020**
    (c) 2021-2030 (d) 2031-2040
13. Which Conservation is mainly used for cultivated plants multiplied by seeds?
    (a) *In-situ* Conservation **(b) *Ex-situ* Conservation**
    (c) Field gene bank (d) Cryopreservation
14. Botanical name of ber is
    (a) *Psidium guajava* (b) *Vitis vinifera*
    (c) *Musa paradisica* **(d) Zizyphus mauritiana**
15. Time of fruit bud differentiation in in aonla is
    (a) March- April (b) February-March
    (c) January February (d) April-May
16. Seeds remain viable for long time at low temperature known as __________ seed.
    (a) Recalcitrant (b) Viable
    (c) Orthodox (d) None of these
17. Beet, turnip and petunia are ______________
    (a) Obligatory short day plants
    (b) Obligatory long day plants
    (c) Facultative short day plants
    (d) Facultative long day plants
18. ________ is phenomenon where development of seed without fertilization occurs.
    (a) Parthenocarpy (b) Sternospermocarpy
    (c) Parthenogenesis (d) None of these

19. Pusa Delicious a gynodioecious variety of papaya has which of the following sex forms.
    (a) Pistillate and Hermaphrodite (b) Pistillate
    (c) Staminate (d) None of these
20. Which of the following types of flowers are present in mango panicles?
    (a) Male and hermaphrodite (b) Female only
    (c) Male and female (d) Male and neutral
21. Pusa delicious and Pusa majesty varieties of papaya are ___________
    (a) Andrmonoecious **(b) Gynodioecious**
    (c) Dioecious (d) None of these
22. _______is a free living aerobic N-fixing bacterium.
    (a) *Azolla* **(b) *Azotobacter***
    (c) *Azospirillum* (d) *Beijerinckia*
23. Frequent pre-monsoon showers facilitate germination of weeds to be destroyed by cultivation is known as____________
    (a) Stale seed bed (b) Raised seed bed
    (c) Flat seed bed (d) None of these
24. Agricultural sector contribution to GDP is
    (a) 15% **(b) 17.9%**
    (c) 18% (d) 19 %
25. State growing mostly tea is
    (a) Maharashtra (b) Uttar Pradesh
    (c) Gujarat **(d) Assam**
26. Movement along the same demand curve shows
    (a) Expansion of demand
    (b) Expansion of supply
    (c) Expansion and contraction of demand
    (d) Increase and decrease of demand
27. Margin share holder of Agriculture Insurance Company is
    (a) National Insurance Company Ltd
    (b) General Insurance Corporation of India
    (c) NABARD
    (d) The New India Assurance Company Ltd.

28. NAIS stand for

    (a) National Agriculture Innovative Scheme

    (b) National Agricultural Infrastructure Scheme

    (c) National Agricultural Insurance Scheme

    (d) National Agricultural and Rural Development Insurance Scheme

29. In case of a straight line demand curve meeting the two axes, the price elasticity of demand at the mid point of the line would be

    (a) 0 **(b) 1**

    (c) 1.5 (d) 2

30. If price is forced to stay below equilibrium price

    (a) Excess supply exists **(b) Excess demand exists**

    (c) Either 1 or 2 (d) Neither 1 or 2

31. Which of the following cost curves is never U shaped?

    (a) Average cost curve (b) Marginal cost curve

    (c) Total curve **(d) Fixed cost curve**

32. The best use of limited resources is studied under the law of

    (a) Substitution (b) Opportunity cost

    (c) Diminishing returns (d) All of the above

33. As on today, the number of nationalized bank in India is

    (a) 25 **(b) 20**

    (c) 21 (d) 14

34. Consider the following statements and identify the correct answer.

    i) The lead bank scheme owes its origin to Nariman Committee

    ii) Under this scheme, SBI and its subsidiaries only were given the responsibility of development of districts

    (a) (i) alone is correct (b) (ii) alone is correct

    (c) Both are correct (d) None

35. The process of .... begins with banks lending money out of primary deposit.

    (a) Credit creation (b) Cash credit

    (c) Debit creation (d) Over draft

36. Consider the following statements and identify the correct statements.

    i) The Kisan Credit Scheme was launched in 2015

    ii) The scheme provides adequate and timely credit support to the farmers for their cultivation needs.

    (a) (i) alone is correct **(b) (ii) alone is correct**

    (c) Both are correct (d) None

37. Physical function of marketing includes

    (a) Grading (b) Buying

    (c) Selling (d) Financing

38. Products which remain untaxed under GST is

    (a) Fresh fruits and vegetables (b) Meat products

    (c) Dairy products **(d) All the above**

39. Who issues Agmark?

    (a) Directorate of Marketing and Inspection, Government of India

    (b) Bureau of Indian Standards (BIS)

    (c) Ministry of Food Processing

    (d) Ministry of Statistics

40. Path coefficient analysis was first used for plant selection by__________.

    (a) Dewey (b) Lu

    (c) Johnson **(d) Both (a) and (b)**

41. If two regression lines coincide, then what is the value of r?

    (a) 0 (b) 1

    (c) ± (d) -1

42. How many phenotypic classes are obtained in a trihybrid cross F_2 population?

    (a) 3 (b) 2

    (c) 16 —' **(d) 8**

43. Each gene of a quantitative character has:_________effect.

    (a) Large and cumulative (b) Unequal

    (c) Small and cumulative (d) Different

44. As the magnitude of environmental effect increases, the phenotypic classes progressively__________.

(a) Overlap each other (b) Separate each other

(c) Become discrete (d) None of these

45. In a random mating population with no epistasis and zero inbreeding, the covariance half sibs is__________.

(a) 0 (b) 1

(c) $\frac{1}{2}\sigma^2 A$ **(d) $\frac{1}{4}\sigma^2 A$**

46. What does a value of one of average degree of dominance indicate n Hayman's graphical approach of diallel analysis?

(a) Over dominance (b) Partial dominance

(c) No dominance **(d) Complete dominance**

47. In L x T analysis, the variance sca is equal to__________.

(a) Covariance full sibs- 2 covariance half sibs

(b) (Cov. FS- Cov. HS)

(c) 2 Covariance full sibs- 2 covariance half sibs

(d) 2 Covariance full sibs- covariance half sibs

48. In Finlay and Wilkinson model, a variety having b value of 1 and low mean yield would indicate a__________genotype.

(a) Most widely adapted **(b) Poorly adapted**

(c) Not adapted (d) None of these

49. In Perkins and Jinks model, GxE sum of squares is subdivided into __________and__________.

(a) Heterogeneity due to regression and SS due to remainder

(b) Genotype and E+ GxE

(c) Genotypes, environments and GxE

(d) Envt(linear), GxE linear and pooled deviations

50. In NCD II, give the formula for variance due to males x females.

(a) $\sigma^2_{mxf} = 1/4V_D$ (b) $\sigma^2_{mxf}=1/2VA+1/4V_D$

(c) $\sigma^2_{mxf} = 1/2V_A$ (d) $\sigma^2_{mxf}=1/2VA+1/2V_D$